T0215656

Design Thinking

This design focused series publishes books aimed at helping Designers, Design Researchers, Developers, and Storytellers understand what's happening on the leading edge of creativity. Today's designers are being asked to invent new paradigms and approaches every day – they need the freshest thinking and techniques. This series challenges creative minds to design bigger.

More information about this series at https://link.springer.com/bookseries/15933

Designing Human-Centric AI Experiences

Applied UX Design for Artificial Intelligence

Akshay Kore

Apress®

Designing Human-Centric AI Experiences: Applied UX Design for Artificial Intelligence

Akshay Kore
Bengaluru, India

ISBN-13 (pbk): 978-1-4842-8087-4   ISBN-13 (electronic): 978-1-4842-8088-1
https://doi.org/10.1007/978-1-4842-8088-1

Managing Director, Apress Media LLC: Welmoed Spahr
Acquisitions Editor: Susan McDermott
Development Editor: James Markham
Coordinating Editor: Jessica Vakili

Distributed to the book trade worldwide by Springer Science+Business Media New York, 233 Spring Street, 6th Floor, New York, NY 10013. Phone 1-800-SPRINGER, fax (201) 348-4505, e-mail orders-ny@springer-sbm.com, or visit www.springeronline.com. Apress Media, LLC is a California LLC and the sole member (owner) is Springer Science + Business Media Finance Inc (SSBM Finance Inc). SSBM Finance Inc is a **Delaware** corporation.

For information on translations, please e-mail booktranslations@springernature.com; for reprint, paperback, or audio rights, please e-mail bookpermissions@springernature.com.

Apress titles may be purchased in bulk for academic, corporate, or promotional use. eBook versions and licenses are also available for most titles. For more information, reference our Print and eBook Bulk Sales web page at http://www.apress.com/bulk-sales.

Any source code or other supplementary material referenced by the author in this book is available to readers on the Github repository: https://github.com/Apress/Designing-Human-Centric-AI-Experiences. For more detailed information, please visit http://www.apress.com/source-code.

Printed on acid-free paper

For
Mummy and Ajji,
*for always loving and supporting me despite
not understanding what I do.*

Kanika,
*for always loving and supporting me
while understanding what I do.*

and our cats, **Momo and Shifu,** *for teaching me
that keyboards make for great cat beds.*

Table of Contents

About the Author

Akshay Kore is a product designer and has led the design of AI products at Fortune 500 companies and high-growth start-ups. He studied interaction design at IIT Bombay.

Akshay has written articles for various publications, like HackerNoon, The Startup, UX Planet, UX Collective, etc., on user experience design, artificial intelligence, and voice interfaces. He frequently talks about designing AI products at conferences, design schools, organizations, and podcasts. Currently, he lives in Bangalore, India, with his wife and two cats.

About the Technical Reviewer

Abhishek Potnis has completed his doctoral degree under the supervision of Prof. Surya Durbha in the GeoComputational Systems and IoT Lab at the Indian Institute of Technology Bombay. His research interests include geospatial knowledge representation and reasoning, natural language processing (NLP), deep learning for computer vision, satellite image processing, remote sensing and GIS, and Internet of Things. His doctoral research explored the areas of geospatial semantics and deep learning for satellite image processing, toward leveraging knowledge graphs for enhanced scene understanding of remote sensing scenes.

Abhishek is an open source enthusiast and has been a code contributor to Mozilla Firefox. When not at his terminal or engrossed in a sci-fi novel, he enjoys traveling, capturing, and captioning the world through his camera.

Acknowledgments

I am incredibly grateful to many people who supported me, gave critical feedback, and helped sharpen the message. I would like to thank Aagam Shah, Aashraya Sachdev, Abhishek Damodara, Ameet Mehta, Ami Dhaliwal, Amrutha Palaniyappan, Anand Vishwaroop, Ankit Kejriwal, Anupam Taneja, Arka Roy, Ayush Kumar, Betson George, Bharadwaj Radhakrishna, Chandra Ramanujan, Dhivya Sriram, Dhruv Shah, Dileep Mohanan, Emi Sato, Isabella Scheier, Ivy Wang, Jayati Bandyopadhyay, Jonathan D'mello, Kanika Kaul, Kaustubh Limaye, Matt Lee, Mitali Bhasin, Nadia Piet, Natalie Abeysena, Neha Singh, Nirvana Laha, Prasad Ghone, Ritwik Dasgupta, Rohan Verma, Sagar Puranik, Saloni Dandavate, Sanket Lingayat, Shilpa Nair, Shirin Khara, Smith Shah, Srayana Sanyal, Sruzan Lolla, Tejo Kiran, and Thommen Lukose.

This book would not have been possible without the support of my editors, Susan McDermott and Jessica Vakili. They helped me resolve countless doubts and kept me on track. I want to thank Natalie Pao for reaching out and providing me with this opportunity. I'm grateful to Dr. Abhishek Potnis for helping me correct technical inconsistencies.

Writers are nothing without readers. Lastly, I am grateful to you for choosing this book and taking the time to read it. Thank you.

Preface

Many of my favorite science fiction authors predict a worldview that I often find inevitable. Humans evolved to be intelligent, and over time, we will imbue intelligence in our things, systems, and environments. In fact, we've already started doing this. If I ask you to find information about "Aryabhata," I am confident you would begin with a Google search. Many of us would prefer our streaming services to suggest things to watch. I used Grammarly, an AI assistant, to correct grammatical mistakes when writing this book. A car that drives itself is just around the corner, and it is not odd to talk to your speaker anymore.

Artificial intelligence is inevitable, and anyone interested in shaping the future should take note. Do an image search for the term *artificial intelligence*. Many results will be abstract representations of brains, networks, and robots. AI starts to feel like an esoteric technology meant for geniuses in labs churning out sophisticated solutions. While parts of this picture are true, AI is also about how we work, play, and live. It is also about making mundane everyday interactions smarter, more efficient, and easy.

Over the last five years, I've had the opportunity to work with AI teams in various large companies and start-ups. I've consumed numerous books, videos, podcasts, articles, and research papers on the topic. I've spoken to over a thousand people from product management, design, engineering, and research. I realized that there is a gap between a designer's understanding of how the technology works and the practical realities of building AI products.

AI brings a fundamental shift to how we design products. Instead of specifying rules, AI teams are responsible for curating outcomes based on algorithms and large amounts of data. AI systems are dynamic, they change over time, and their user experience needs to adapt to this change. This calls for a shift in how we think about designing intelligent products.

Designing good AI products is also a highly collaborative process between different disciplines. It is extremely unlikely that a lone designer will be responsible for building and shipping AI. A big part of being a good AI designer is being a good team member.

Over time, I started compiling my learnings on what works and what doesn't when building AI products and how designers can become effective and impactful team members. This book is merely a distillation of this collected knowledge.

Who Should Read This Book

You should read this book if you are a designer, technologist, researcher, manager, or founder working with or interested in building human-centered AI products. No prior experience with AI is required to read this book. This book can help you identify opportunities for using artificial intelligence to solve problems and create desirable AI products. The contents of this book are also relevant for developers interested in building intuitive interfaces for AI-based solutions.

Overview of Chapters

This book is divided into four parts:

1. Part 1 (Intelligence) comprises Chapters 1 and 2. We discuss the idea of intelligence, the many meanings of AI, and the view of artificial intelligence this book subscribes to.

2. Part 2 (Decisions) consists of Chapter 3, which talks about making decisions regarding incorporating AI in your workflows and finding opportunities for using it in your organization.

3. Part 3 (Design) focuses on specific design patterns, techniques, and ethical considerations for designing human-centric AI products. This part consists of Chapters 4–8.

4. Part 4 (Teamwork) describes how designers can communicate and collaborate effectively with AI tech teams and consists of Chapters 9 and 10. This section introduces essential AI terminologies, behaviors, and mindsets to help you become an impactful and effective team member.

How to Use This Book

I wrote this book as a practical guide to designing human-centric AI products. And even though I've written this book, I'm pretty sure I will have to keep going back to refer to sections from time to time. The pace at which we forget things is astonishing.

While you might want to finish it in one go, in most cases, you will encounter different types of challenges at different times in your product development journey. I would recommend that you read it once and then use the book as a reference whenever you get stuck or when you want to explore different approaches to solve a problem with AI. Think of it as a working handbook for designing AI products, a tool in your design arsenal.

I sincerely hope that you find the contents of this book useful, and I wish you all the best.

Akshay Kore

PART 1

Intelligence

CHAPTER 1

On Intelligence

This chapter looks at the many meanings of artificial intelligence and its foundations. We talk about the point of view of AI that this book subscribes to and discuss why artificial intelligence needs human-centered design.

I think my cat understands me. Need pets? She meows. Hungry? She meows. Want treats? She meows. She has her goals—food, treats, or pets—and knows how to achieve them. Although I have no way of knowing if she understands concepts like "goals" or "achievement," she gets what she wants, for the most part. I think I would describe her as intelligent.

The futurist Max Tegmark describes intelligence as the ability to achieve complex goals.[1] By that definition, my cat does some remarkable things. She knows whom to ask for pets or food and how to ask for it. She has different approaches for achieving different outcomes. For food, her strategy is to look me in the eye and meow real loud. For pets, she'll quietly snuggle next to me and utter a soft meow. Smart.

You and I can do a lot more. If we want food, most of us can make something from what's available in the kitchen. We can also do other things like work in teams, negotiate, navigate unknown spaces, build stuff, or write code. Groups of people can run companies and even entire nations—something almost impossible for a single person to do. Humans can collaborate across cultures and generations. Getting affectionate pets, on the other hand, is difficult for people.

---

[1] Tegmark, Max. *Life 3.0*. Penguin Books, 2018.

© Akshay Kore 2022
A. Kore, *Designing Human-Centric AI Experiences*,
https://doi.org/10.1007/978-1-4842-8088-1_1

Many Meanings of AI

Intelligence is a strategic advantage. Everything humanity has to offer is a product of intelligence. Building artificial intelligence could be the biggest event in human history.[2] But most people working in AI can't seem to agree on a standard definition for the word. Intelligence is what Marvin Minsky calls a suitcase word, packed with different meanings like cognition, consciousness, emotion, etc. The term *artificial intelligence*, coined by John McCarthy in the 1950s, inherits this problem. However, some people say that this lack of a precise definition might be a good thing. It has enabled the field to advance faster and wider.[3]

In his widely acclaimed book *Artificial Intelligence: A Modern Approach*, Stuart Russel lays out multiple definitions for AI.[4] I've added a few more that fit the model. The vital question to ask when defining AI is whether you are concerned with thinking or behavior.

---

[2] Agrawal, Ajay, et al. *Prediction Machines*. Harvard Business Review Press, 2018.

[3] Mitchell, Melanie. *Artificial Intelligence*. First ed., Farrar, Straus and Giroux, 2019.

[4] Russell, Stuart J., and Peter Norvig. *Artificial Intelligence*. Third ed., Pearson, 2016.

Table 1-1.   *Some definitions of AI. The top two rows are concerned with the thought process, while the bottom two are concerned with behavior*

Thinking humanly	**Thinking rationally**
1. "The exciting new effort to make computers think…machines with minds, in the full and literal sense."[5]	1. "The study of mental faculties through the use of computational models."[7]
2. "[The automation of activities] that we associate with human thinking, activities such as decision making, problem solving, learning …"[6]	2. "The study of the computations that make it possible to perceive, reason, and act."[8]
	3. Artificial intelligence is the overarching science that is concerned with intelligent algorithms, whether or not they learn from data.[9]

(*continued*)

---

[5] Haugeland, John. *Artificial Intelligence*. A Bradford Book—The MIT Press, 1989.

[6] Bellman, Richard. *An Introduction to Artificial Intelligence*. Boyd & Fraser Pub. Co., 1978.

[7] Charniak, Eugene, and Drew McDermott. *Introduction to Artificial Intelligence*. Addison-Wesley, 1987.

[8] Winston, Patrick Henry. *Artificial Intelligence*. Addison-Wesley, 1993.

[9] Husain, Amir. *The Sentient Machine*. Scribner, 2018.

Table 1-1. (*continued*)

Acting humanly	Acting rationally
1. "The art of creating machines that perform functions that require intelligence when performed by people."[10]	1. "Computational intelligence is the study of the design of intelligent agents."[14]
2. "The study of how to make computers do things at which, at the moment, people are better."[11]	2. "AI…is concerned with intelligent behaviour in artifacts."[15]
3. "We think of AI as a set of technologies that enable computers to perceive, learn, reason and assist in decision-making to solve problems in ways that are similar to what people do. "[12]	3. "Intelligence, defined as the ability to accomplish complex goals, can't be measured by a single IQ, only by an ability spectrum across all goals."[16]
4. "Systems that extend human capability by sensing, comprehending, acting, and learning. Machine-learning application as one that builds models based on data sets that engineers or specialists use to train the system."[13]	4. "Artificial intelligence has been mainly about working out the details of how to build rational machines."[17]

---

[10] Kurzweil, Ray, and Diane Jaroch. *The Age of Intelligent Machines*. MIT Press, 1990.

[11] Rich, Elaine, et al. *Artificial Intelligence and the Humanities*. Third ed., McGraw Hill Education, 2017.

[12] Smith, Brad, and Harry Shum. *The Future Computed*. Microsoft Corporation, 2018.

[13] Wilson, H. James, and Paul R. Daugherty. *Human + Machine: Reimagining Work in the Age of AI*. Harvard Business Review, 2018.

[14] Poole, David, et al. *Computational Intelligence: A Logical Approach*. Oxford University Press, 1998.

[15] Nilsson, Nils J. Artificial Intelligence. Morgan Kaufmann, 1998.

[16] Tegmark, Max. *Life 3.0*. Penguin Books, 2018.

[17] Russell, Stuart. *Human Compatible*. Allen Lane, an imprint of Penguin Books, 2019.

Thinking Humanly

Definitions of AI that focus on thinking like humans follow a cognitive modeling approach. If we want to build a machine that thinks like humans, we need to determine a way of how humans think. These approaches focus on what goes inside the human mind. There are three key ways[18] of understanding this:

1. Through introspection by catching our thoughts

2. Through psychological experiments by observing a person in action

3. Through brain imaging by observing the brain in action

This approach focuses on building a sufficiently precise theory of the mind and translating that into a computer program.

Thinking Rationally

In these definitions of AI, there is an attempt to codify rational thinking. Aristotle was the first person to do this through his idea of syllogisms. A syllogism provides a pattern for structuring an argument that always yields the correct result given the context. These structures are supposed to be irrefutable. "Socrates is a man; all men are mortal; therefore, Socrates is mortal" is a famous syllogism.

---

[18] Russell, Stuart J., and Peter Norvig. *Artificial Intelligence*. Third ed., Pearson, 2016.

Nineteenth-century logicians took this idea one step further by defining precise notations about statements, objects, and their relationships. Computational reasoning systems started appearing by 1965 that, in principle, could solve any logically solvable problem. If not, the program looped forever.

This approach hopes to build intelligent systems on the foundations of logic. However, we run into two critical problems:

1. The logical approach assumes 100% certainty about knowledge of the world. In the real world, informal knowledge is often uncertain and incomplete, and in general, no single person or agent has complete understanding.

2. There are currently limits to computation. There is a big difference between solving a problem in theory and solving it. Even a few hundred parameters, in theory, can exhaust computational resources. The real world has a lot more.

Acting Humanly

If a computer can fool us into thinking it is human, it is acting humanly and is intelligent. This is a popular definition and is also sometimes referred to as the Turing test approach. Trying to fool humans has led to significant advances in adjacent technologies in AI. The computer needs the following capabilities to act humanly:[19]

1. **Natural language processing** to understand and communicate with people

---

[19] Russell, Stuart J., and Peter Norvig. *Artificial Intelligence*. Third ed., Pearson, 2016.

2. **Knowledge representation** to store and organize what it knows and hears

3. **Automated reasoning** to use the stored information to answer questions and carry out conversations

4. **Machine learning** (ML) to find patterns and adapt to changing circumstances

5. **Computer vision** to perceive objects

6. **Robotics** to move around and manipulate objects

All of these are hard problems and compose most of modern AI.

Acting Rationally

This approach is also called the rational agent approach. An agent is something that acts. An intelligent agent can do more; it can operate autonomously, perceive its environment, create plans, and pursue goals. A rational agent acts to achieve the best outcome or the best-expected outcome when there is uncertainty.

In the "thinking rationally" approach, the focus is on correct inferences. This aligns with "acting rationally" since making the correct inferences is essential for achieving the best possible outcome. However, sometimes decisions need to be made even when there is no provably right thing to do. For example, a reflex action like recoiling after touching a hot iron is an action without inference or rationale.

Likewise, to "act humanly" like in the Turing test approach also involves skills that allow an agent to "act rationally." Knowledge representation and reasoning enable an agent to act rationally. We need to be able to communicate well to act rationally.

However, achieving perfect rationality—doing the right thing all of the time—is often not feasible given incomplete information and uncertainty in environments. Rational agents, in many cases, need to act appropriately with limited information or computational resources. This is called "limited rationality," which is how most real-world AI systems operate.

The rational agent approach has two advantages over other approaches:[20]

1. It is more general than the "thinking rationally" approach because correct inferences are just one way of acting rationally.

2. It is more amenable to scientific development than approaches based on human behavior or thinking. Rationality is mathematically well defined and completely general, and we can unpack it to generate AI designs that achieve goal-oriented behavior.

Note   This book adopts the view that intelligence is concerned with acting rationally. An intelligent agent takes the best possible action in a situation despite limited information or resources. We will look at designing AI experiences that are intelligent in this sense.

---

[20] Russell, Stuart J., and Peter Norvig. *Artificial Intelligence*. Third ed., Pearson, 2016.

Substrate Independence

You can consider many things to be intelligent using the rational agent definition. You and I are intelligent; my cat is intelligent, and so is a robotic vacuum cleaner that cleans the dust off floors. We all exhibit goal-oriented, intelligent behavior in uncertain environments. Out of the preceding examples, three are biological, and one is silicon hardware–based, yet they all feel intelligent.

The quality of intelligence has an intangible, ethereal, or abstract feel because it is substrate independent: it can take a life of its own irrespective of the details of the underlying material substrate.[21] Here are some important points to note about substrate independence:

1. Substrate independence doesn't mean that a substrate is unnecessary. It means that the details of the underlying substrate are irrelevant. Intelligence can exist in networks of neurons in the brain as well as on silicon-based computers.

2. We are primarily interested in the substrate-independent aspect of intelligence and AI. Although modern AI is closely associated with computer science that works with silicon hardware, if we successfully build AI on biological substrates, we won't abandon designing AI systems since the underlying substrate does not matter.

---

[21] Tegmark, Max. *Life 3.0*. Penguin Books, 2018.

Figure 1-1.   Substrate independence. The cats and the robotic vacuum cleaner both feel intelligent. Source: Photo by YoonJae Baik on Unsplash

Foundations of Artificial Intelligence

Although modern AI is closely related to computer science, many disciplines have pondered over different aspects of building intelligent agents. Some of these fields have contributed significant viewpoints and techniques to the domain of AI. The following are some of the foundational disciplines:

1. Philosophy

2. Mathematics

3. Economics

4. Neuroscience

5. Psychology

6. Computer engineering

7. Control theory and cybernetics

8. Linguistics

9. Business

Philosophy

Philosophy deals with questions that are important to building intelligent machines, like "How does knowledge lead to action?", "How does a mind arise from a mushy brain?", or "Can formal rules be used to draw valid conclusions?" Philosophers going back to 400 BC have considered the idea of the mind as a machine, that it operates on knowledge encoded in some internal language. At the same time, the machine can use thoughts to choose the appropriate action. Aristotle's syllogisms discussed earlier have been instrumental in the field of knowledge representation.

Intelligent behavior requires reasoning to justify actions. Parts of philosophy deal with the connection of knowledge and action, that is, reasoning, which is vital to building rational agents since intelligence requires action backed by proper justifications.

Mathematics

Philosophers asked, "Can we use formal rules to draw valid conclusions?" while mathematicians asked, "How can formal rules be used to derive valid conclusions?" The leap from philosophy to formal science required mathematical formalization in three fundamental areas: logic, probability, and computation.[22] Mathematicians provided tools and techniques to manipulate logical, probabilistic, or uncertain statements. Mathematics also sets the foundation for computation and reasoning through algorithms.

Economics

Economics is not really about money. Most economists study how people make choices that lead to preferred outcomes. Economists formalized the problem of making decisions that maximize the expected outcome for the decision-maker.[23] In AI, a rational agent needs to make appropriate decisions in uncertain situations. The field has borrowed many ideas from economics involving single or multiple agents in an environment.

Neuroscience

Neuroscience is the study of the nervous system and the brain. Neuroscientists discovered many facts about how brains work and how similar or different they are to or from computers. Many AI techniques like perceptrons and neural networks got their start by emulating the brain through learnings from neuroscience.

---

[22] Russell, Stuart J., and Peter Norvig. *Artificial Intelligence*. Third ed., Pearson, 2016.

[23] Russell, Stuart J., and Peter Norvig. *Artificial Intelligence*. Third ed., Pearson, 2016.

Psychology

Cognitive psychology views the brain as an information processing device. Thereby, we can consider humans and animals as information processing machines. Different movements in psychology have tried to understand how humans think and act, and artificial intelligence has used many of its learnings to build important AI components. One such example is in the field of knowledge representation. In his book *The Nature of Explanation*, Kenneth Craik specified three critical steps for a knowledge-based agent to translate a stimulus into an internal mental model.[24] Cognitive processes manipulate this internal model to derive a new model or knowledge representation used to take action. Most intelligent organisms build a small-scale internal model of external reality to navigate the world and make decisions. They run mental simulations, and as more information is acquired, the model is tweaked. This mechanism helps the organism better adapt to a changing environment.

Computer Engineering

The fields of AI and computer engineering have a symbiotic relationship. While we discussed that intelligence is substrate independent, much of modern AI has chosen the computer as its substrate of choice. Computer science provided the field with powerful computing capabilities that make AI applications possible, while work in AI has pioneered many valuable ideas to computer science like time-sharing, windows and mice, linked lists, and object-oriented programming.

---

[24] Craik, Kenneth. *The Nature of Explanation*. Cambridge University Press, 1967.

Control Theory and Cybernetics

The control theory and cybernetics fields focus on designing machines that act optimally based on feedback from the environment. Although the tools used by control theory are different from those of AI, both fields have a similar goal of operating in environments by making decisions in uncertain situations. Making such decisions involves learning, which happens through feedback from the agent's environment.

Linguistics

Linguistics is the scientific study of knowledge. Understanding language requires more than simply understanding sentences and grammar. It involves knowing the context and understanding the subject matter. The problem of understanding language is considerably more complex than it looks. Modern linguistics and AI gave rise to a hybrid field called natural language processing, or NLP. NLP deals with the interaction between humans and computers using natural language. Much of the early work in knowledge representation, that is, the study of knowledge put into a form that a computer can reason, was informed by research in linguistics.[25]

Business

AI will be helpful wherever intelligence is valuable. The field would not have progressed so much without a business case for AI's economic value. Governments and organizations have invested large sums in AI research, and modern AI has already generated billions of dollars in value.

---

[25] Russell, Stuart J., and Peter Norvig. *Artificial Intelligence*. Third ed., Pearson, 2016.

Why Is AI a Separate Field?

The foundations described previously don't necessarily focus on building intelligence as their ultimate goal. AI borrows tools and techniques from these disciplines. It lies at their intersection. The following are a few reasons for AI to branch out into a separate field:

1. AI is unique in attempting to duplicate mental faculties like creativity, learning, self-improvement, and language use.

2. Unlike other disciplines discussed previously, AI is the only field that attempts to build intelligent machines that function autonomously in complex, uncertain environments.

3. Because of its affinity to computer science, the field is amenable to the scientific process of hypothesis building and experimentation.

Superintelligence and Artificial General Intelligence

Those working in AI frequently encounter fears of people outside the field guided largely by popular depictions of sentient machines and superintelligent AIs (SAIs) taking over the world. Superintelligent AIs (SAIs) are systems whose intelligence far surpasses humans in every way. Researchers warn about the threats of such systems to society and are working toward finding ways of controlling such systems. Many others say that such systems are far out in the future, and there is no point worrying right now. However, most agree that superintelligence is not an immediate threat. We should be wary of over-allocating resources to it.

Before building superintelligence, we will have to build AI that is equivalent to human intelligence. Known as artificial general intelligence (AGI), such a system can learn and perform any intellectual task that a human can.

Note   Since these types of AIs are nowhere on the horizon, designing experiences for SAIs and AGIs is not in scope for this book.

Narrow AI

When you think of modern AI, think of an algorithm, not a sentient robot. Narrow artificial intelligence is the ability to achieve specific goals in an environment. Much of the value that modern AI has produced comes from doing specific tasks extremely well, like converting speech to text (STT), recognizing objects or people from images, or creating a playlist tailored just for you. A smart speaker like Amazon's Alexa, Google Home, or Apple's Siri comprises multiple AI algorithms that do individual tasks. For example, they can detect a wake word like "Alexa" or "Hey, Siri" or convert what you say into a command like speech to text or text to action, among many others.

Most AI companies and products operate in the narrow AI domain. There are high chances that any AI system you may have encountered is specialized to accomplish a specific task. A self-driving car can drive well but cannot fetch you coffee (yet). This book focuses on designing AI products primarily in the field of narrow AI.

Despite being called "narrow," modern AI has generated significant value with some important applications. Some of these are listed below:

- *Face recognition*: The ability to recognize faces from an image is helpful in many applications like sorting photos by people in them or unlocking your phone with your face.

- *Object detection*: Detecting objects from images helps self-driving cars or a robotic vacuum cleaner navigate environments.

- *Robotics*: AI algorithms allow robots to do motion planning and control motors to follow a path or perform an action like picking up an object.

- *Spam detection*: Many email spam filtering programs use AI techniques like text classification to separate spam and non-spam messages.

- *Web search*: Many popular search engines like Google use AI to surface the most relevant results using AI techniques like knowledge representation and ranking algorithms.

- *Speech to text*: This is the ability to convert spoken words into text widely used in speech recognition software and virtual AI assistants like Alexa or Siri.

- *Machine translation*: Tools like Google Translate use machine translation to translate between languages automatically.

Figure 1-2. Examples of narrow AI. *(a) A robotic arm is picking up objects. Source: Photo by David Levêque on Unsplash. (b) Google's self-driving car. Source: Wikimedia Foundation. (c) AI-generated playlists on Spotify. Source: Photo by David Švihovec on Unsplash. (d) Amazon Alexa is responding to a voice query. Source: Photo by Lazar Gugleta on Unsplash. (e) Face ID on Apple iPhone. Source: Photo by David Švihovec on Unsplash. (f) Google web search. Source: Photo by Edho Pratama on Unsplash. (g) Amazon product recommendations. (h) Navigation on Google Maps. Source: Photo by Isaac Mehegan on Unsplash*

Rules vs. Examples

So far, you've seen that artificial intelligence is a broad field that borrows tools and techniques from several disciplines. We can use it in a wide array of applications ranging from spam detection to robotics. However, when building AI systems, there are two fundamental approaches: based on rules or examples.

Rules-Based

In the rules-based approach, the AI is programmed to follow a set of specific instructions. The algorithm tells the computer precisely what steps to take to solve a problem or reach a goal. For example, a robot designed to navigate a warehouse is given specific instructions to turn when there is an obstruction in its path.

Examples-Based

Also known as sub-symbolic AI, in this approach, the AI is taught, not programmed. Learning happens by giving the AI a set of examples of what it will encounter in the real world and how to respond. For example, to detect faces from an image, an AI may be shown multiple images of faces in different environments. By observing these images, the AI learns how to detect a face.

A significant amount of progress in AI has been in the examples-based approach of machine learning (ML). This type of approach is advantageous when it is difficult to specify rules for a task (like detecting text from speech, recommending a video, generating a painting). By learning from examples, the AI builds a model of how to accomplish a goal without specifying exact steps.

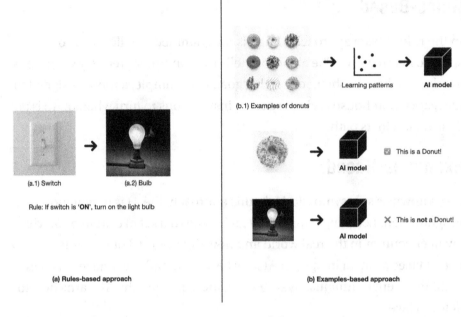

Figure 1-3. Approaches of building AI. (a) Rules-based approach. Here, the switch is mapped to the light bulb. The rule given is to turn on the light bulb by flicking the switch on. (b) Examples-based approach. The AI learns to recognize donuts by showing it examples of what a donut looks like without specifying exact steps

A Fundamental Difference in Building Products

While some AI systems are rules-based, many new AI-based products are built on top of the examples-based approach of machine learning (ML). So much so that AI in the industry is sometimes synonymous with ML. Machine learning brings a fundamental change to how we build software products. Instead of programming the system to do a specified action, its creators provide data and nurture it to curate desired outcomes. ML is handy when it is difficult to specify exact rules. For example, if you had to build a system that detects donuts, you might start with a basic definition of the food: a fried cake of sweetened dough in the shape of a ring that people eat. This definition contains many terms for which it would be tough to program exact rules. A computer has no concept of a cake, of sweetness, or of what eating means. It does not know what dough is or what frying means. We feel that computers understand the meaning; this is an illusion. In the case of machine learning, it is irrelevant to know the meaning of terms. You can show the computer pictures of donuts, and it will build a model for recognizing them.

When building modern AI products, the role of creators is to curate lots of examples or data to help the system create a model of what you want it to do. Creators are training the model. The model will not always be perfect. The job of creators then is to give it feedback so that the AI improves over time.

Intelligent Everything

Andrew Ng, the former head of Google Brain, calls AI the "new electricity." It is the idea that AI will soon become as pervasive and invisible as electricity itself. Like we discussed earlier, intelligence is substrate independent. It is also a generic property, which means that anything can become intelligent. Your smartphone can be intelligent, so can your car, and so can your coffee cup. I don't argue for making everything smart, but it is a possibility. With the reducing costs and improving speed of computing, networks, and data storage, it could be feasible to make every object intelligent.

A database is another example of a technology that is invisible and pervasive. The majority of modern applications, like ecommerce, food delivery, flight bookings, etc., use databases. Just like the database got inside all software, AI will get inside all software.[26] And just like we don't call our food delivery applications "database products," we will stop calling applications that include intelligent components "AI products."

User Experience for AI

Artificial intelligence is a transformative technology. AI is substrate independent, and potentially any object can become intelligent. We can use it in a vast number of applications, from robotics to automatic language translation. However, we've placed a lot of focus on the technical aspects of AI by trying to define the term and explaining how it works and its foundations and much less on its human element.

Beneficial AI

Because we build most AI applications and products for humans, AI needs to work for you and me. It needs to be beneficial for us. AI needs human-centered design. At one level, AI will require that even more people specialize in digital skills and data science.[27] However, building beneficial AI products will require more than science, technology, engineering, and math skills. As we deploy AI in society, disciplines like social sciences, humanities, ethics, economics, psychology, and philosophy will be critical in helping manage and develop beneficial AI. While AI has impacted and will continue to positively impact society, designing beneficial AI products presents its own problems and unique challenges:

---

[26] a16z. AI: What's Working, What's Not. 2017, www.youtube.com/watch?v=od7quAx9nMw.

[27] Smith, Brad, and Harry Shum. *The Future Computed*. Microsoft Corporation, 2018.

1. First is the **control problem**, which is how to build AI
    that is beneficial to us. How can we know cases when
    it is not working in our favor? What tools or techniques
    can we use to make sure that AI does not harm us?

2. The second is **building trust**. For AI to be helpful,
    its users need to trust it. We need to make sure that
    users of AI can trust its results and make appropriate
    choices. For example, the majority of the results of
    an AI system are probabilistic. When it predicts a
    donut in an image, it is never 100% certain. It could
    be 99.9% certain but never 100%. There is always a
    margin for error.

3. The third significant problem is **explainability**.
    Many AI systems are notorious for operating in a
    black box: the reasons for its suggestions are not
    known or are difficult to explain. The problem of
    explainability deals with providing appropriate
    information or explanations for AI's results so that
    its users can make informed judgments.

4. **Ethics** is a critical ingredient for designing
    beneficial AI products. Our societies are biased,
    and many times AI models reflect this underlying
    bias, harming users or leading to unintended
    consequences. AI ethics focuses on formulating
    values, principles, and techniques to guide moral
    conduct when building or deploying AI.

AI is an important and consequential technology. UX designers,
product designers, and managers are advocates for users and stakeholders.
Our job as product creators is to ensure that the AI products we build
are beneficial to our users. I believe we can achieve this by focusing on a
human-centered process when designing AI products.

Summary

This chapter defined AI and described its foundations and the need for human-centered design when building AI products. Here are some of the important points:

1. There are many definitions of artificial intelligence. This book subscribes to the rational agent view of AI. A rational agent acts to achieve the best outcome or the best-expected outcome when there is uncertainty.

2. Intelligence is a substrate-independent property. It can exist irrespective of the underlying material substrate. Intelligence can be present in biological forms like humans as well as silicon forms like computers.

3. Although modern AI is closely related to computer science, it has its foundations in several disciplines: philosophy, mathematics, economics, neuroscience, psychology, computer engineering, control theory and cybernetics, linguistics, and business.

4. We will mainly discuss AI from the point of view of narrow AI—an ability to do specific tasks extremely well. Discussions around artificial general intelligence (AGI) and superintelligence are beyond the scope of this book.

5. Intelligence is also a generic property, that is, anything can become intelligent.

6. For AI to work, it needs to be beneficial for the people it impacts. We need human-centered design to build beneficial AI products.

CHAPTER 2

Intelligent Agents

In the previous chapter, we subscribed to the rational agent definition of AI. In this chapter, we define rational agents in an environment. We describe how AI works from the lens of input-output mappings, feedback loops, and rewards. Toward the end of the chapter, we discuss the probabilistic nature of AI models.

Of all the players in a football (or soccer) team, I find the role of the goalkeeper the most interesting. While the rest of the team focuses on scoring, the goalkeeper's responsibility is to ensure that the other team does not. Their literal job is to keep goals from happening. Although it takes a lot of physical and mental skill to become a good goalkeeper, their job can be distilled into this simple step—if the ball is entering the goalpost, don't let it.

I've left out several details in this oversimplification, like as follows: the action of blocking a goal needs to happen on a football field, the goalkeeper needs to be a member of a playing team, and they need to be at the correct goalpost.

In this case, the rational action for the goalkeeper is to block the football from entering the goalpost. The football field, the goalpost, players, and the football comprise the environment where this action occurs, and the goalkeeper is the one that acts.

© Akshay Kore 2022
A. Kore, *Designing Human-Centric AI Experiences*,
https://doi.org/10.1007/978-1-4842-8088-1_2

Rational Agent

The idea of the intelligent or rational agent is a central concept of modern AI.[1] An agent is something that acts. A rational agent acts so as to achieve the best outcome or, when there is uncertainty, the best possible outcome.[2] We expect intelligent agents like computers to do much more than just react to stimuli—they need to operate autonomously, perceive their environments, adapt to change, create goals, and pursue them.

Achieving perfect rationality, that is, doing the right thing all the time, is not always feasible, especially in changing, uncertain, or complex environments. Most real-world environments are complex, and there is insufficient time or computing power available. Therefore, most rational agents hope to achieve the best possible outcome by making the best possible decisions. They operate within what is known as limited rationality.

Agents and Environments

An agent acts within an environment, as well as on it. In the previous example, the agent is the goalkeeper who acts in the environment of the football field. The football field environment comprises players, goalposts, the football, referees, and the field itself. The goalkeeper uses their eyes to perceive whether a ball is coming their way. If the football moves in their direction, they use their hands, legs, or entire body to block it. The goalkeeper is the agent that perceives the environment of the football field through sensors like the eyes and acts on the incoming football with actuators like limbs. Let's break this down.

---

[1] Russell, Stuart. *Human Compatible*. Allen Lane, an imprint of Penguin Books, 2019.

[2] Russell, Stuart J., and Peter Norvig. *Artificial Intelligence*. Third ed., Pearson, 2016.

Agent

An agent is anything that perceives its environment through sensors and acts on it. The agent does this action through actuators. A robotic vacuum cleaner is an agent that acts on the environment of a room to clean it. Humans are agents that perceive through sensors like eyes, ears, nose, tongue, or touch. We act through actuators like hands and legs.

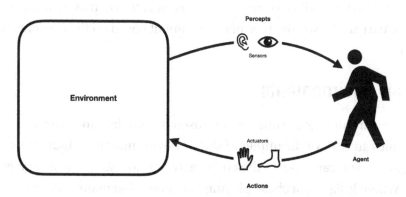

Figure 2-1. *An agent acts with its environment through sensors and actuators. Humans are agents that perceive through sensors like eyes, ears, nose, tongue, or touch. We act through actuators like hands and legs*

Environment

An environment is what an agent acts on. A room or a house is the environment for a robotic vacuum cleaner. For a chess-playing machine, its environment is the chessboard.

The design of an agent depends on the type of environment and the nature of the problem. You can also think of an environment as the problem statement and the rational agent as its solution. For example, in a chess-playing machine, the problem statement or environment is to win against a player by ending in a checkmate within the rules of chess and the board's space. The algorithm (or steps) that the machine uses is the solution to the problem of beating the opponent in chess.

Not all environments are the same. A chessboard is a very different environment from a dusty house. Environments can have multiple agents. A chessboard has two agents; a robotic vacuum cleaner is a single agent in an environment. Most real-world problems like coaching a football team, building a product, or running an organization consist of multiple agents. We can classify the type of environment based on the complexity of the task.

Simple Environments

Simple environments are observable, discrete, and deterministic, and their rules are known. Examples of simple environments include puzzles and games. You can observe the entire state of a chessboard at any given time. When designing a chess-playing machine, what matters is the current state of the board, not what happened before. Chess has known rules and is reasonably predictable by a machine. Building a chess-playing machine is an easy problem, and AI researchers have developed fairly general and effective algorithms and a solid theoretical understanding.[3] Machines often exceed human performance on such problems in simple environments.

Complex Environments

Most real-world environments are complex. Running an organization and teaching astrophysics are hard problems. They have complex, mostly unobservable environments, multiple agents and objects, different types of agents and entities, no specific rules, long timescales, and lots of

---

[3] Russell, Stuart. *Human Compatible*. Allen Lane, an imprint of Penguin Books, 2019.

uncertainty. We have ways of solving parts of complex problems but no general method to solve them completely. In the real world, most complex problems are broken down into simple problems. These simple problems are then tackled individually and sometimes simultaneously.

Sensors

An agent uses sensors to perceive its environment. We use the eyes, ears, nose, and other organs to sense the environment. A robotic vacuum cleaner might use cameras, infrared, and other sensors to navigate a room and its obstacles. A software that corrects grammar might use keystrokes, words, or phrases as sensors within the environment of words and grammar rules.

Actuators

Actuators are anything the agent uses to act on the environment. Our goalkeeper uses hands, legs, and other body parts as actuators to block a football. A self-driving car might use its engine, tires, lights, and other instruments as actuators.

Goals

Agents act on their environment with a particular objective. The simplest way to communicate an objective is in the form of goals. These goals may be static like winning a chess game or dynamic like finding the best possible route where the definition of best possible may change depending on various factors like the time it takes to reach, cost efficiency, or road conditions.

Table 2-1. *Examples of agents and environments with their sensors, actuators, and goals*

Goal	Agent	Environment	Sensors	Actuators
Clean a room	Robotic vacuum cleaner	Room, dust	Camera, infrared	Vacuum suction, brush
Find something on the Internet	Search engine	Websites on the Internet	Keyboard input	Display of search results
Sort vegetables	Vegetable sorting machine	Pile of vegetables	Camera	Robotic arms
Email spam detection	Spam detection algorithm	Emails in the inbox	Incoming email, other user information	Mark as spam
Convert speech to text	Speech-to-text engine	Audio file	Microphone	Display of text transcription
Predict the price of a house	Price estimation algorithm	Specifications of houses like location, size, number of rooms, price of similar houses, etc.	Keyboard input or mouse click on a listing	Display of estimated price

Input-Output

In modern AI, intelligence is closely related to pattern recognition. The idea of mapping inputs to outputs is another way of thinking about AI. For example, a rational agent like a robotic vacuum cleaner receives input about dust particles through sensors like cameras, infrared, etc. and acts on it to produce an output of a clean room. For the robot, seeing dust on the floor results in the action of removing it. What it sees is the input; a clean floor is an output.

You and I are great at recognizing patterns. We learn to match faces to identify people. To speak, we match the sound to the correct meaning. To read, we match squiggly lines on paper to a word. Generally, being smart is about knowing lots of patterns, that is, matching input to output.[4] A good chess player has memorized lots of opening, middle, and endgame patterns. Doctors know how to match symptoms to the correct diagnosis and cure. A good designer knows how to match user needs to the appropriate solution.

Similarly, an AI system tries to match a given input to the desired output. For example, speech-to-text engines match the input of a sound file into a word, or face recognition systems map images to people's faces. You can use this simple idea of mapping input and output to build some very innovative and valuable products. Spam filters can match the input of an email to whether it is spam or not. In factories, we can use images of a product to detect defects. Online advertising systems use user information to predict a person's likelihood of clicking an ad. Input from cameras, infrared, radar, and other sensors can tell a self-driving car the position of people and objects.

---

[4] Polson, Nicholas G, and James Scott. *AIQ*. Bantam Press, 2018.

Table 2-2. *Examples of input-output mappings in AI*

Input	Output
Image of a place  *Image source: Rowan Heuvel on Unsplash*	Place: "Taj Mahal" Location: "Agra, India"
Audio file	Transcription: "I am a cat"
Image search: "Cat"	 *Image source: Cédric VT on Unsplash*

(*continued*)

Table 2-2. (*continued*)

Input	Output
	Image classification: "Donut / Not Donut"
Quote: "An eye for an eye only ends up making the whole world blind."	Who said it: "Mahatma Gandhi"
"Bud'dhimattā"	Language: "Marathi" Translation: "Intelligence"

Learning Input-Output Mappings

In the previous chapter, we looked at two key approaches of building AI, namely, rules-based and examples-based. In the rules-based approach, the AI is provided with a set of instructions to achieve an outcome. In the examples-based approach, instead of specifying steps, the AI learns input-output mappings through seeing examples. The examples-based approach is especially useful when it is difficult to describe exact steps for a problem. For example, we show a robotic vacuum cleaner lots of images of clean and dirty surfaces so that it learns when to clean a surface or leave it as is.

Machine Learning (ML)

Machine learning or ML is a type of AI technique that uses the examples-based approach. With machine learning, the system can accomplish tasks without giving any specific instructions. The AI system starts forming patterns that help it match input to the desired output by showing it lots of examples. We are essentially teaching machines how to learn. It builds a mathematical function for this input-output mapping. We call this mathematical function the ML model. To teach a machine, we provide it with lots of data. We divide this data into a training and test set. The kind of data provided depends on the goals of the AI system you are building. For example, to create a face recognition system, you might give the machine lots of pictures of faces. In general, there are three primary steps of teaching machines how to learn, namely, training, testing, and deployment:

1. *Training*: In this step, we show the machine data from the training set. The machine learns from this data by building patterns for desired input-output relationships like detecting images with faces in them. The output of this step is an ML model.

2. *Testing*: We use a test set to validate the accuracy of the ML model generated during training. If the model performance is not good enough, we tweak some model parameters and train it again. We repeat this loop of training and testing till we achieve the desired performance.

3. *Deployment*: After the model is trained and tested for adequate performance, we deploy it into the world.

While the steps are generally similar, there are different approaches for teaching machines how to learn. The following are a few popular methods.

Supervised Learning

In supervised learning, we use labeled data to build the model. *Labeled* means that along with the example, we also provide the expected output. For instance, we can label a face recognition dataset by indicating whether an example image contains a face or not. If we want to build a system that detects product defects in a factory, we can show the ML model many defective and good-condition products. We can label these examples as "defective" or "good condition" for the model to learn.

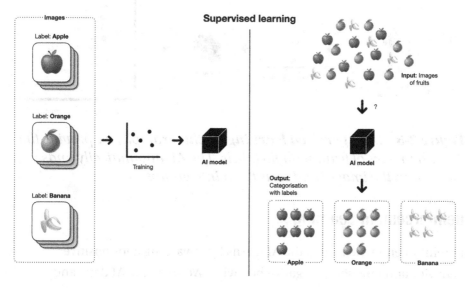

***Figure 2-2. Supervised learning.** In this example, we train the AI with labeled data of images of fruits. After training, the AI model can categorize fruit images by their labels*

Unsupervised Learning

We provide no labeled data in unsupervised learning. The AI learns by finding patterns in the input data on its own. Most unsupervised learning focuses on clustering—grouping data by some characteristics or features.[5] These features or groups are then used as labels that the model generates. This technique is especially useful when you want to detect anomalies in data, for example, credit card fraud, where specific behavior patterns may be observed, or grouping a set of images by types like people, animals, landscapes, etc.

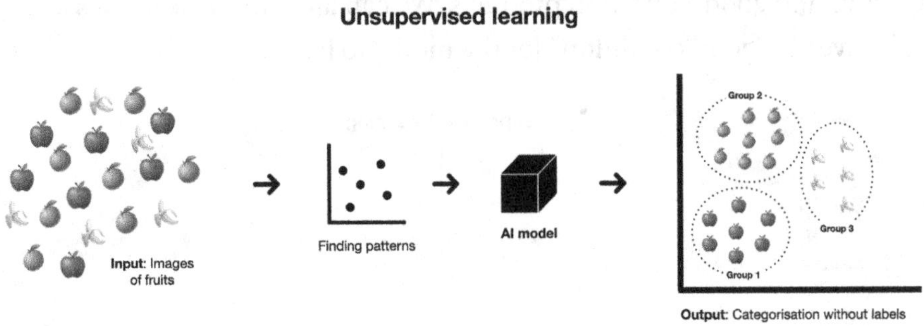

Figure 2-3. Unsupervised learning. In this example, we provide the AI with images of fruits with no labels. The AI automatically finds patterns in the images and sorts them into groups

Reinforcement Learning

In reinforcement learning, the AI learns by rewarding it for positive behavior and punishing negative behavior. We show the AI data, and whenever it produces correct output, we reward it (sometimes in the form of points). When it produces incorrect results, we punish it by not rewarding it or deducting rewards. Over time the AI builds a model to get the maximum reward. Training a pet is a good analogy for understanding reinforcement learning. When the pet displays good behavior, you give

[5] "IBM Design for AI." ibm.com, www.ibm.com/design/ai/.

them treats, and when not, you don't give them any treats. Over time the pet learns that if they behave in a particular manner, they will get treats, and you get good behavior as an outcome.

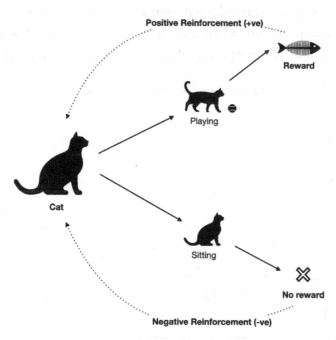

Reinforcement learning

Figure 2-4. Training a cat using positive and negative reinforcement

Deep Learning (DL)

Deep learning (DL) is a subset of machine learning. In ML, learning happens by building a map of input to output. This map is the ML model represented as a mathematical function. The mapping can be straightforward or convoluted and is also called a neural network. For example, clicking a button to switch on a light bulb is a simple neural network. The input is the click, and the output is the state of the light bulb (on/off). Learning this sort of mapping is easier and sometimes referred to as shallow learning. For more complex cases like predicting the price of a house based on input parameters like size, the number of rooms, location, distance from schools, etc., learning happens in multiple steps. Each step is a layer in the neural network. Many of these steps are not known and are called hidden layers. Learning with multiple layers is known as deep learning. In ML, depth refers to the number of layers in the neural network. Networks with more than three layers are generally considered deep. DL is uniquely suited to build AIs that often appear human-like or creative, like restoring black-and-white images, creating art, writing poetry, playing video games, or driving a car.

Deep learning and Neural Networks

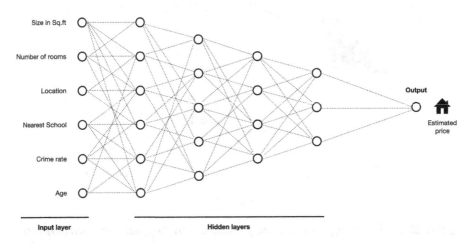

Figure 2-5. Deep learning and neural networks*. In this example, features of a house like its size, number of rooms, location, etc. are used to estimate its price*

Feedback Loops

Learning anything comprises two steps—consuming knowledge and correcting mistakes through feedback. Machine learning models change or improve over time through a continuous feedback mechanism. The feedback loops help the ML model correct errors or fill gaps in its information.

Rewards

How do we know that a feedback loop has resulted in an improvement? In ML, we do this through a reward function. The reward function is loosely based on a training technique in psychology called operant conditioning. We use operant conditioning to train animals by giving them rewards for good behavior and punishment for bad behavior.

Also known as the loss function, the reward function is a central concept of machine learning. Like training a pet, the reward function allows the ML model to receive positive or negative feedback. Based on the feedback, the ML model optimizes itself to receive the maximum rewards and improve over time.

A reward function is not generated automatically. Product teams carefully design the reward function to achieve particular outcomes or sets of behaviors. You get the results that you design for. Designing the reward function is often a collaborative process between product management, engineering, machine learning, and design, among other disciplines.

The Risk of Rewards

Machines do not have an intrinsic desire to chase rewards. We provide them with objectives in the form of goals. When they achieve the desired outcome, they receive positive feedback or a reward. ML models optimize themselves to achieve the maximum possible reward and thereby our desired objectives. Stuart Russel refers to this as the standard model of building AI. Most AI applications are built on the foundation of this standard model.

However, there is a downside to blindly following the standard model. The risk is not that machines won't do their jobs well. The risk is in machines doing it too well. Worse, if we put the wrong objectives, this leads to unintended consequences that may harm users and society. AI recommendation systems on social media optimize themselves for maximum user engagement. Optimizing for user engagement leads to user-specific recommendations that subtly change user behavior over time, sometimes leading to bias, polarization, and threatening to upend existing social structures. Imbuing AI with objectives imperfectly aligned to ours is the problem of value alignment.

AI systems optimize to achieve our preferences. However, human preferences are always uncertain, constantly changing. It is impossible to define true human preference.[6] It turns out this is a feature, not a bug. To build AI that is beneficial to us, we need intelligence that satisfies our preferences even if they are uncertain. The AI needs to understand the difference between what is good for humans and what achieves its objectives. This is a hard problem to solve.

The Probabilistic Nature of AI

When you think of AI, think of an algorithm, not a robot. We know that machine learning models improve over time, which means that they are not perfect; there is always some scope for improvement. Unlike traditional software, where outcomes are deterministic, that is, you click a button to send an email or press the shutter to take a picture, ML systems are probabilistic. The outputs of ML models are never 100% certain. They could be 99.9% sure, but never 100%. There is always a margin of error.

Machine learning models are "soft": they don't follow strict logic. For a speech-to-text engine, "write" and "right" sound the same but have different meanings depending on the context. Maybe a bald person's head looks like a duck egg from a certain angle for an object recognition system. We can better grasp the workings of ML models if we embrace this "softness" and understand how these algorithms make decisions and mistakes. Having this understanding can help in building better AI products.

---

[6] Russell, Stuart. *Human Compatible*. Allen Lane, an imprint of Penguin Books, 2019.

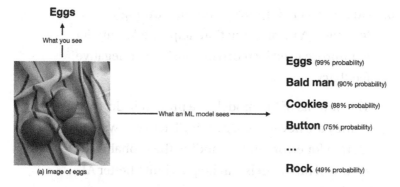

Figure 2-6. Probabilistic nature of AI. You see eggs, while the AI sees eggs, a bald man, cookies, etc., with varying probabilities

Summary

This chapter looked at the concept of agents and environments, input-output mappings, and machine learning. Here's a summary:

1. The idea of the rational agent is a central concept of modern AI. A rational agent acts to achieve the best outcome or, when there is uncertainty, the best possible outcome.[7]

2. An agent is anything that perceives its environment through sensors and acts on it through actuators.

3. Intelligence is related to pattern recognition— matching an input to the desired output.

4. Machine learning (ML) is an AI technique that allows a computer to learn input-output mappings without giving it specific instructions. The AI learns input-output mappings by seeing lots of examples.

---

[7] Russell, Stuart J., and Peter Norvig. *Artificial Intelligence*. Third ed., Pearson, 2016.

5. An ML system improves over time through feedback. A reward function helps the ML model to improve itself by receiving positive or negative feedback.

6. The outputs of ML models are probabilistic, that is, they are never 100% certain. There is always a margin for error. Understanding the probabilistic nature of ML models can help us build better AI products.

PART 2

Decisions

CHAPTER 3

Incorporating Artificial Intelligence

This chapter looks at the differences between humans and machines and how they complement each other. We will understand how you can incorporate AI in your workflows and find opportunities for using AI in your organization.

Most of us would figure out how to make a cup of instant coffee in a typical kitchen. If there's a coffee machine, you'd find the coffee, add water, find a mug, and brew it by pressing the right buttons. If there's no machine, you'd probably look for coffee, water, sugar, and milk to brew in a container on a cooktop. Then you'd pour it in a mug. It's a pretty easy job for most people.

If we put our mind to it, assembling an IKEA table is another job that most of us could do. You'd figure out instructions from the manual, understand which piece fits where, and build the table. Sometimes you'd have to improvise, but I trust you'd figure it out.

We are yet to see an AI that can brew a cup of coffee from the kitchen or assemble an IKEA table, let alone do both. Easy things are hard.

© Akshay Kore 2022
A. Kore, *Designing Human-Centric AI Experiences*,
https://doi.org/10.1007/978-1-4842-8088-1_3

A Cognitive Division of Labor

Adam Smith's idea of allocating roles based on relative strengths in the eighteenth century mainly focused on dividing physical labor. Smith suggested that individual employees working on single subtasks would increase productivity. Each person can specialize in a task and delegate their weaknesses to another.

Computers, software, and AI enable a cognitive division of labor based on the machine and human's cognitive strengths.[1] Humans and machines can overcome their individual weaknesses and focus on their strengths.

What Machines Do Better

1. Unlike humans, machines do not get tired and can operate nonstop. For example, a mobile banking application does not need to take lunch breaks.

2. Machines can perform routine, repetitive, and redundant tasks much faster than people. For example, a single bottling machine can fill hundreds of bottles simultaneously in the time it takes a human to fill just one.

3. Multiple machines can share information instantaneously, which means that when one machine learns something new, all of them can learn. For example, you can install a software update on all machines at the same time.

---

[1] Agrawal, Ajay, et al. *Prediction Machines*. Harvard Business Review Press, 2018.

4. Machines can also distribute learning into chunks of information. A hundred robots solving different parts of a problem in parallel for ten hours equals one thousand hours of learning.

5. Machines in the form of software can create millions of copies of themselves quickly without significant costs.

6. A machine's intelligence, that is, its software, can be decoupled from its hardware. For example, you can access and use email from any computer with an Internet connection, and using the email software is not dependent on the device. On the other hand, you can't separate a human body from its brain (yet).

7. Machines can analyze vast amounts of data faster than people, which you can use to generate predictions or detect anomalies.

What Humans Do Better

1. While machines are good at churning large amounts of data, people are good at interpreting ambiguous and complex information that might be present.

2. Humans are better at operating in uncertain and complex environments like typical kitchens, governments, or companies. Unlike most machines, people can adapt on the fly.

3. Humans have better physical dexterity in certain situations like conducting surgery or manipulating wires.

4. While an AI prediction can reduce uncertainty, human judgment is needed to make a decision. We need judgment in determining the payoff or utility of a particular outcome. AI is good for making routine decisions, but human judgment in decision-making is critical for rare events.

5. You can trust people more than machines in making moral and ethical choices for other humans.

6. Humans can learn from small amounts of data quickly. For example, you can teach a child what an apple looks like from just a few examples.

7. People can learn concepts from a particular situation and apply them in a different one. This is known as transfer learning. For example, learning to drive a car can help in learning to drive a truck faster.

8. People are better at interpersonal skills than machines. People are better at interpreting social and emotional cues, and they can collaborate more effectively.

We often think of humans and machines as adversaries fighting for each other's jobs. But we neglect powerful opportunities in collaboration between the two sides. Machines are good at doing routine tasks quickly, sifting through large amounts of data to generate predictions or detect anomalies. At the same time, humans are better at interpreting complex information and making judgments with limited information. Humans and machines can be symbiotic partners that help and improve each other leading to better outcomes.

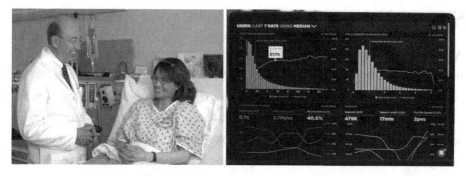

Figure 3-1. Cognitive division of labor. (Left) Humans are better at jobs like nursing that involve interpersonal skills. Source: Photo by the National Cancer Institute on Unsplash. (Right) Machines are better at churning large amounts of data like analytics. Source: Photo by Luke Chesser on Unsplash

> *Humans and machines can be symbiotic partners that help and improve each other.*

Human + Machine

Most jobs that people do are a combination of tasks. For example, writing a report involves multiple tasks like researching, organizing content, creating a narrative, editing, etc. Computers and word processing software help in writing content, while search engines help find information. An organization is essentially lots of people doing lots of jobs and tasks along with machines. Machines improve the speed or quality of an activity, and sometimes machines work alongside people. It is rare to find an organization with only people or only machines in it.

Supermind

A complex organization like a company or a government sometimes feels intelligent, like it has a mind of its own. We frequently hear things like "Company X thinks about privacy and security." We personify organizations. Humans and machines can act together in an organization that can seem intelligent from the outside. We looked at the idea of intelligence being substrate independent in the previous chapter. In the case of an organization, the underlying substrate would be people, machines, processes, and their interconnections. Thomas Malone refers to this idea as the Supermind—a group of individuals acting together in ways that seem intelligent.[2] The "individual" in this case could be a person or a machine.

An organization is a type of a Supermind whose goal is to achieve the best possible outcome—thereby acting intelligently.

Collective Intelligence

We usually think of people or machines as intelligent. An intelligent agent tries to achieve the best possible outcome in an environment, which means different things for different agents. For example, a clean floor is the best outcome for a robotic vacuum cleaner, while maximum profits would be the best outcome for a stock trading algorithm.

It is fair to assume that most large projects are done by groups of people and machines, often within organizations. Most organizations have goals, and achieving them is their best possible outcome. An organization is a type of a Supermind whose goal is to achieve the best possible outcome—thereby acting intelligently. We call this collective intelligence— the result of a group of individuals acting together in ways that seem

---

[2] Malone, Thomas W. *Superminds*. Oneworld, 2018.

intelligent.[3] Collective intelligence is something that makes organizations effective, productive, adaptable, and resilient.[4]

Even if their employees or departments might not have the same goals, all organizations work toward achieving a particular outcome. For example, the goal of a company could be maximizing profits or shareholder value, while the purpose of its marketing department could be to improve engagement and leads. These are not the same goals, but they work toward achieving the desired outcome.

You can think of this desired outcome as the organization's impact, which can be increased by improving its effectiveness and productivity or making the organization more adaptable or resilient. We can improve the organization's impact by increasing its collective intelligence. There are three ways of increasing collective intelligence:

1. Addition

2. Improvement

3. Connection

Addition

You could increase collective intelligence by adding more people or machines into the organization.

1. Adding more people generally refers to hiring additional employees. More people can lead to increased capacity and output and make an organization resilient.

---

[3] Malone, Thomas W. *Superminds*. Oneworld, 2018.
[4] Kore, Akshay. "Systems, Superminds and Collective Intelligence." Wok Toss Robot, 2020, https://woktossrobot.com/2020/05/12/systems-superminds-and-collective-intelligence/.

2. Adding more machines could mean purchasing additional equipment or software, leading to improvements in capacity, output, efficiency, and speed of reaching desired outcomes.

Improvement

You could increase collective intelligence by improving machines or people in the organization.

1. Improving machines generally means upgrading hardware or software, which leads to better efficiency, capacity, output, and speed of reaching desired outcomes.

2. People improve themselves through learning and feedback by upskilling, reading, or studying something new. Learning generally leads to improved productivity and efficiency for the organization.

Connection

Improving collaboration between people, machines, or both can increase the collective intelligence of an organization. We often underestimate the power of hyperconnectivity and strengthening connections between humans and machines. Groups of people and machines are far more collectively intelligent than their individual parts.

1. **Improving human-human connections**

There are different ways of improving the connection between people. For example, we can think of incorporating better communication tools, restructuring the organization, organizing group activities, or even making changes to the workplace to improve collaboration. These interventions often lead to greater collaboration, increase in adaptability, or resilience in organizations.

2. **Improving machine-machine connections**

    Improving the connection between machines generally means streamlining workflows—for example, improving computational power, network speeds, writing APIs, enhancing security, software that automates tasks for a program, etc. Enhancing machine-machine connection generally leads to improvements in capacity and resilience.

3. **Improving human-machine connections**

    These types of interventions are prevalent in an organization. Connection improvements between people and machines are of two types:

    a. Machines help people in doing their tasks. For example, a word processing program helps write reports efficiently, or a system alerts a factory supervisor if a particular part is not functioning correctly. These generally lead to improved productivity and efficiency.

    b. People help machines by writing software updates and correcting the machine's mistakes, which often leads to improved quality and speed of outcomes.

Collective intelligence—the result of a group of individuals acting together in ways that seem intelligent.[5]

Artificial Collective Intelligence

When we decide to incorporate artificial intelligence into an organization, we are essentially trying to increase its collective intelligence. You can think of AI as an element that permeates the workings of an organization—sometimes used as a tool to add new functionality like search or, sometimes, as a material that changes the structure of jobs like prediction algorithms that suggest actions to workers. AI can help people and machines augment their tasks or sometimes even completely automate them. There are four ways in which AI can increase the collective intelligence of an organization:

1. Improving machines

2. Automating redundant tasks

3. Improving machine-machine collaboration

4. Improving human-machine collaboration

Improving Machines

You can think of AI as an ingredient that improves existing software or hardware. Improvements can range from the increased efficiency of the machine, making it more humane, or enabling new types of experiences. An update to a music streaming service can enable new AI experiences like personalized playlists. Automobile manufacturers can upgrade cars to have enhanced safety features or new capabilities like autopilot without changing their hardware.

---

[5] Malone, Thomas W. *Superminds*. Oneworld, 2018.

Automating Redundant Tasks

Searching for a photo or an article manually requires a lot of time and effort. Sorting large volumes of items like fruits manually is tedious. AI can free up time, creativity, and human capital by automating low-value, routine, and redundant tasks, improving the organization's overall productivity and thereby its collective intelligence.

Improving Machine-Machine Collaboration

Most machines in homes or factories operate in silos and are focused on completing a single task. They do not communicate with other machines in their environment. Like we saw earlier, there are benefits to hyperconnectivity and strengthening connections between components. You can improve collective intelligence by strengthening ties. AI can help machines in communicating and collaborating with each other. For example, a smart home hub can enable communication and collaboration between all the lights in the house to achieve the perfect lighting for a situation or save energy costs.

Improving Human-Machine Collaboration

In a factory, machines often perform specific tasks away from humans. Humans control these machines with simple mechanisms like pressing buttons, adjusting levers, etc. We look at machines as mere tools to use. With AI, people and machines collaborate with each other as symbiotic partners in a system. Imagine a nimbler robot that works alongside humans to perform repetitive tasks. People are adaptable and learn quickly. If a new type of task comes in, humans can teach the robot by showing it how to do it.[6] Overnight the robot has a new capability, and

---

[6] Knight, Will. "This Factory Robot Learns a New Job Overnight." MIT Technology Review, 2016, www.technologyreview.com/2016/03/18/161519/this-factory-robot-learns-a-new-job-overnight/.

people can focus on other tasks. Improving human-machine collaboration can increase the organization's collective intelligence, thereby improving productivity and efficiency.

> *You can think of incorporating AI as a way of increasing the collective intelligence of the organization and thereby improving its impact.*

Missing Middle

Companies that are using machines merely to replace humans will eventually stall, whereas those that think of innovative ways for machines to augment humans will become the leaders of their industries.[7] Teams of humans and AI are more collectively intelligent than them working in silos. Teams of humans plus machines can dominate even the strongest computers. The chess machine Hydra, which was a chess-specific supercomputer like Deep Blue based in the United Arab Emirates, was no match for a strong human player using an ordinary computer. Human strategic guidance combined with the tactical acuity of a computer was overwhelming.[8]

Humans and machines can work together to form powerful synergies. This is the missing middle. In the missing middle, humans and machines aren't adversaries, fighting for each other's jobs. Instead, they are symbiotic partners, each pushing the other to higher levels of performance.[9] With this approach, people improve the AI, while AI gives them superpowers. Companies can let go of rigid processes for greater flexibility and achieve more significant outcomes.

---

[7] Wilson, H. James, and Paul R. Daugherty. *Human + Machine: Reimagining Work in the Age of AI*. Harvard Business Review, 2018.

[8] Kasparov, Garry, and Mig Greengard. *Deep Thinking*. John Murray, 2018.

[9] Wilson, H. James, and Paul R. Daugherty. *Human + Machine: Reimagining Work in the Age of AI*. Harvard Business Review, 2018.

Humans and machines can work together to form powerful synergies.

Cobots

Some robots can even collaborate with people. Cobots or collaborative robots are flexible machines that work with people as partners. In the case of factories, this can mean moving away from rigid assembly lines in favor of organic, flexible, and adaptable teams of humans and advanced AI systems. We also refer to this idea as organic AI. As BMW and Mercedes-Benz have experienced, rigid assembly lines are giving way to flexible teams of employees working closely alongside robots. Moreover, these novel types of teams can continuously adapt on the fly to new data and market conditions. With these numbers, you might expect a well-oiled, robot-dominated assembly line operating with as few people as possible. But Mercedes is ditching some of its robots and redesigning its processes to center them around people. The automotive assembly line is changing, and the driver of this change is the rise of customizable cars. You can now go online and choose from an expansive array of features on your next car. Now, to fulfill customized orders and handle fluctuations in demand, employees can partner with robots to perform new tasks without manually overhauling any processes or manufacturing steps. Those changes are baked into the system and are performed automatically. Instead of designing an assembly line to make one kind of car at a time, these lines can adapt as needed.[10]

You can think of AI as a member of your organization and assign it specific roles.

---

[10] Wilsom, H. James, and Paul R. Daugherty. *Human + Machine: Reimagining Work in the Age of AI*. Harvard Business Review, 2018.

Roles for AI

When incorporating AI into your workflows, you can also think of AI as a part of the organization and assign it specific roles. You can give AI different levels of control, and these roles can range from tools, to assistants, and to AI managing a group of people.

Tools

When people use AI as a tool, they have the most amount of control. Search functionality that helps you find text in images is an example of AI as a tool.

Assistants

Although the line between a tool and an assistant is thin, you can think of assistants as systems that respond to your requests and are proactive at the same time. Grammarly, a software that corrects your grammar while you write, and Google Assistant proactively suggesting when you should leave for work are examples of assistants.

Peers

You can think of AI as a peer when it completes an independent function without the involvement of other humans. A robot in a factory assembly line acts like a peer working alongside workers. It performs the assigned job, which is not controlled by the human next to it.

Managers

AI as a manager is generally a system that organizes people and sometimes assigns them tasks to achieve greater output. Although the idea of an AI manager may sound scary, we are already surrounded by many machine managers. Think of a traffic light, air traffic control systems, or the automated management of an Amazon warehouse.[11]

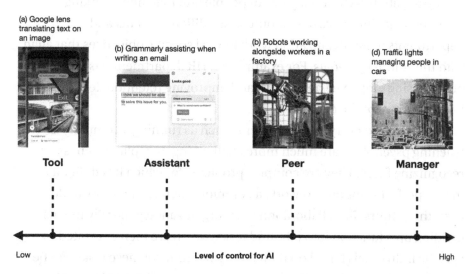

Figure 3-2. Roles for AI. The level of control that AI has is lower on the left and higher on the right. (a) AI as a tool: Google Lens translating text on images. Source: https://play.google.com/ store/apps/details?id=com.google.ar.lens. (b) AI as an assistant: Grammarly assisting when writing an email. Source: www. grammarly.com. (c) AI as a peer: robots working alongside people in a factory. Source: Photo by Lenny Kuhne on Unsplash. (d) AI as a manager: traffic lights managing cars and people. Source: Photo by Florian Wehde on Unsplash

> *Some tasks can be done better by people, while some can be performed better by an AI.*

---

[11] Dzieza, Josh. "Robots Aren't Taking Our Jobs—They're Becoming Our Bosses." The Verge, 2020, www.theverge.com/2020/2/27/21155254/ automation-robots-unemployment-jobs-vs-human-google-amazon.

Finding AI Opportunities

An organization is essentially lots of people and machines working together to achieve a particular outcome—different teams and departments in an organization work toward specific objectives that align with the company's goals. For example, an HR department has the goal of increasing employee headcount and improving retention, leading to higher output for the company.

AI cannot do everything. Problems such as running a company or teaching psychology are much more challenging than playing chess or recognizing faces. They are complex problems for which it is difficult to find perfect solutions. At work, a common theme is not to try to do everything yourself. Collaboration in an organization generally means dividing up a function into jobs and tasks, which are then delegated to specific individuals based on their strengths. Some of these tasks can be done better by people, while some can be performed better by an AI.

Whenever you employ people, add machines, or include AI, think of it as hiring the person, machine, or the AI to accomplish your goal.

Jobs

A job is a way of accomplishing a goal. Organizations have many jobs to be done. Whenever a company employs people, adds machines, or includes AI, they essentially hire the person or AI to accomplish a goal. Any job is a collection of tasks performed by an individual or a team. For example, the job of an architect is to design a building. Within this job, the architect does multiple tasks like talking to clients, creating drawings, planning space, choosing materials, etc. The job of the recruitment team is to find the right candidate. Systems can have jobs too. The job of the digestive system is to digest food. Similarly, machines also have jobs, for example, the refrigerator has a job of keeping food cold.

Using AI can lead to four implications for jobs:[12]

1. AI can augment jobs, as in the case of AI assistants that correct grammar when writing an email.

2. AI can contract jobs by making parts of the job shorter, as in the case of robots in fulfillment centers.

3. AI can lead to a reorganization of jobs. For example, radiologists might spend more time on analysis than sifting through scans.

4. AI can shift the emphasis of specific skills required for a job. For example, in the case of self-driving school buses, the driver may act more like a coordinator, emphasizing communication skills over purely driving ability.

Tasks

A task is a unit of activity that an individual performs to do their job. A job is a collection of multiple tasks. For example, a product designer aims to achieve the best possible user experience given the constraints. They accomplish this by performing various jobs like talking to users, building personas, wireframing, prototyping, communicating with stakeholders, etc. You can further break these jobs into individual tasks like sending emails, planning timelines, choosing suitable interaction patterns, ensuring accessibility, etc.

---

[12] Agrawal, Ajay, et al. *Prediction Machines*. Harvard Business Review Press, 2018.

A study by the McKinsey Global Institute identified that it is technically feasible to automate 50% of activities—not jobs, but tasks—in the US economy with AI. If we map these tasks back to jobs, they found that about 60% of occupations have a third of their tasks automatable.[13] This tells us that many more jobs will be complemented rather than replaced by AI in the near future.

Many more jobs will be complemented rather than replaced by AI in the near future.

Most real-world problems are complex. Organizations operate in mostly unobservable environments with a great deal of uncertainty over long timescales. We are yet to build AI that can run an organization effectively. However, AI can help people with some of their tasks, enhancing overall effectiveness. Incorporating AI can improve the speed and quality of outcomes by streamlining processes, automating tasks, speeding up people's jobs, etc. You can think of incorporating AI as a way of increasing the collective intelligence of the organization and thereby improving its impact. But using AI comes with its tradeoffs. It doesn't always make sense to use AI for all tasks. In this section, we look at how to identify opportunities for AI in an organization. We can divide this process into a few steps:

1. Breaking down jobs into tasks

2. Mapping user journeys

3. Checking if AI is suitable for the task

4. Checking the feasibility of using AI for the task

5. Introducing humans in the loop

---

[13] Ford, Martin R. *Architects of Intelligence*. Packt, 2018.

Breaking Down Jobs into Tasks

A critical part of finding AI opportunities is to identify jobs that need to be done. The next step is to break these jobs down into individual tasks. The goal of this exercise is to break down a job into manageable chunks. Let's understand this through an example.

Example: Personal Running Trainer

Let's say my goal is to run a full marathon. To achieve this goal, I need to be able to run further with consistency. I hire a personal trainer to help me with my training. The job of a running trainer is to improve your running. Improvement can be in the form of speed of the run, distance traveled, or how long you can run. We can break down this job into the following tasks:

1. **Create a training plan.** The trainer needs to create a training plan for me, which can be a personalized or a default beginner plan.

2. **Identify the right equipment.** To run the marathon efficiently, I need to find the right shoes. The trainer can help me with this.

3. **Provide motivation.** There will be days when I will not want to get out of bed. The trainer will need to push me to train.

4. **Ensure I stick to the plan by reminding me.** I might forget workouts or their timings. The trainer can remind me about upcoming workouts.

5. **Tell me my progress.** The trainer might decide to give me regular feedback on my performance.

6. The trainer could also help update the schedule if I miss a session or show me good routes to run.

You can decide to break these tasks down even further or choose to include additional subtasks. The size of these tasks would depend on the level of detail you need.

Mapping User Journeys

Once you have identified jobs and broken them down into manageable tasks, the next step is to create user journeys. These are also called task flows or workflows. It is a map of how the organization operates by representing the user experience and interactions between stakeholders. It need not be extremely accurate but should create a narrative from the user's perspective. Mapping user journeys and existing workflows is a great way of finding opportunities for using AI. As you walk through the user journey, you can better understand tasks in which AI can help, that is, augment, and those you can delegate to an AI, that is, automate. We will discuss augmentation and automation in the next section of this chapter.

When building a map of workflows, it is essential to understand where you would be incorporating AI. Are you integrating AI into an existing system or creating an entirely new workflow? An analysis by the McKinsey Global Institute on 400 use cases across 19 industries shows that more than two-thirds of the opportunities to use AI are in improving existing systems.[14] About 16% of the AI use cases need entirely new workflows. These are also known as greenfield scenarios and are frequent in start-up situations or industries with lots of rich data like healthcare, logistics, or customer service. The analysis tells us that you can find most of the opportunities for AI in existing workflows and processes.

---

[14] Chui, Michael, et al. www.mckinsey.com/, 2018, www.mckinsey.com/featured-insights/artificial-intelligence/notes-from-the-ai-frontier-applications-and-value-of-deep-learning. Accessed July 19, 2021.

In more than two-thirds of our use cases, artificial intelligence (AI) can improve performance beyond that provided by other analytics techniques.

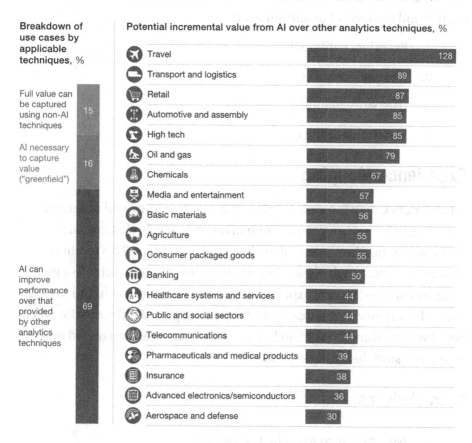

Breakdown of use cases by applicable techniques, %

Potential incremental value from AI over other analytics techniques, %

✈	Travel	128
🚚	Transport and logistics	89
🛒	Retail	87
🚗	Automotive and assembly	85
🏢	High tech	85
⛽	Oil and gas	79
⚗	Chemicals	67
🎬	Media and entertainment	57
🏭	Basic materials	56
🐄	Agriculture	55
🥫	Consumer packaged goods	55
🏛	Banking	50
🩺	Healthcare systems and services	44
🤝	Public and social sectors	44
📡	Telecommunications	44
💊	Pharmaceuticals and medical products	39
📋	Insurance	38
🔲	Advanced electronics/semiconductors	36
⚙	Aerospace and defense	30

Full value can be captured using non-AI techniques — 15

AI necessary to capture value ("greenfield") — 16

AI can improve performance over that provided by other analytics techniques — 69

McKinsey&Company | **Source:** McKinsey Global Institute analysis

Figure 3-3. Opportunities for using AI across industries. Source: McKinsey Global Institute. URL: www.mckinsey.com/featured-insights/artificial-intelligence/notes-from-the-ai-frontier-applications-and-value-of-deep-learning

You can find most of the opportunities for using AI in existing workflows and processes.

There are different methods of mapping user journeys, of which the following are the most relevant. Each can be used depending on your context and the level of detail you need.

1. Experience mapping

2. Journey mapping

3. User story mapping

4. Service blueprints

Experience Mapping

An experience map is a visualization of an entire end-to-end experience that a "generic" person goes through in order to accomplish a goal.[15] This visualization is agnostic of the business or product. This method aims to understand what users experience when trying to achieve a goal. Once you understand the experience, you can tailor your solution to user needs. Experience maps are especially great for greenfield scenarios when building something from scratch. Experience maps can be created even before you start designing a solution.

Characteristics

1. Not specific to a product or service.

2. Events are depicted chronologically.

3. The journey is split into four slices: phases, actions, thoughts, emotions. When finding AI opportunities, you can think of some phases as jobs and some actions as tasks.

---

[15] Gibbons, Sarah. "UX Mapping Methods Compared: A Cheat Sheet." Nielsen Norman Group, 2017, www.nngroup.com/articles/ux-mapping-cheat-sheet/.

4. They are used to generate a baseline understanding of how users accomplish their goals currently.

5. Useful for generating a broad understanding of the customer experience.

6. Suitable for rethinking a product or greenfield scenarios where you are building a product from scratch.

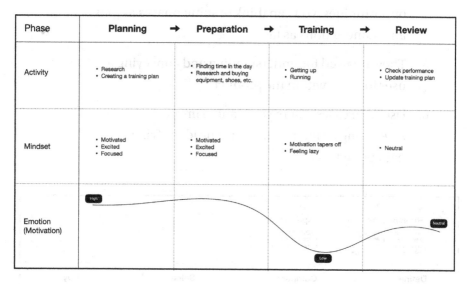

Figure 3-4. *Experience map for a person training for a marathon*

Journey Mapping

A customer journey map is a visualization of the steps a user goes through within a product or a service to achieve a goal. This method combines storytelling and visualization. To build a journey map, you need to conduct primary and secondary research through user interviews, market research, contextual inquiry, direct observation, etc.

Characteristics

1. The map is tied to a specific product or service.

2. It is specific to a user persona.

3. Events are depicted chronologically.

4. The map is split into four parts: phases, touchpoints, actions, and emotions. When finding AI opportunities, you can think of some phases as jobs and some actions as tasks.

5. They are used for understanding and conveying a user journey within the product.

6. Useful in cases where you want to improve an existing product or a service by identifying pain points.

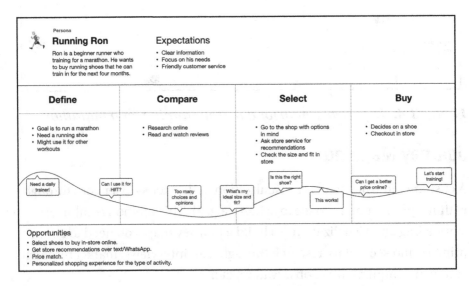

Figure 3-5. *Journey map for a person buying shoes in a store*

User Story Mapping

A user story map is an informal natural language description of the features of your system and is a widely used tool in product management and software development. It consists of multiple user stories in a narrative or storyboard. Think of it as a comic strip of how different stakeholders might use your product. A storyboard is not a specification; it is meant to be a backdrop for discussion with your development team.

Characteristics

1. Although the storyboard may or may not be tied to a specific product, individual stories narrate how a product may be used.

2. A story is specific to a persona. A storyboard may include multiple personas.

3. Events are depicted chronologically.

4. The map is split into two parts: a topline narrative known as an epic and user stories. User stories are arranged according to their priority under an epic.

5. You can think of a user story as a task and an epic as a job.

6. Useful for conveying the user journey and aligning the team.

7. User story mapping is beneficial when building a new product or a feature.

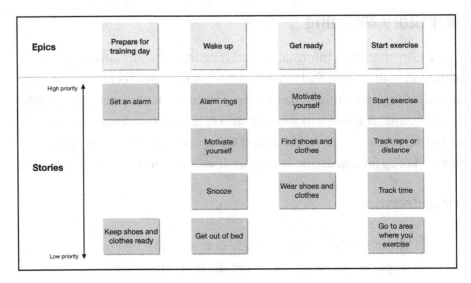

Figure 3-6. *User story map for a person planning to exercise daily*

Service Blueprints

A service blueprint is a detailed visualization of the relationships and interactions between people, machines, artifacts, and processes. It helps build a comprehensive picture of an organization or service.

Characteristics

1. It is tied to a specific product or service.

2. It is specific to a user persona.

3. It includes multiple stakeholders and their relationships.

4. Events are depicted chronologically and hierarchically.

5. The blueprint is split into five sections: user journey, frontstage actions (actions by stakeholders directly interacting with the user), backstage actions (actions by stakeholders not directly interacting with the user), artifacts, and processes.

6. You can think of different frontstage and backstage actions as tasks.

7. You can use it to discover weaknesses and optimization opportunities, bridge cross-department efforts, and break down silos in the organization.

8. It is suitable for finding opportunities in a complex existing system.

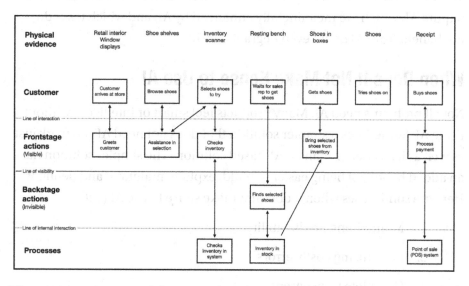

Figure 3-7. *A service blueprint for a running shoe store*

AI won't help you figure out what problems you need to solve.

Problem-First Approach

An essential part of being an AI person is identifying problems that need to and can be solved with AI. Many teams and companies jump right into product strategies that start with AI as the solution while skipping over finding a meaningful problem to solve. Jumping to solutions can lead to the team spending a lot of effort and resources toward addressing a very small or even nonexistent issue.

AI won't help you figure out what problems you need to solve. To do that, you need to go through the effort of finding the right problems through user interviews, ethnography, contextual inquiries, reading customer support tickets, product logs, metrics, etc. Instead of asking "Can we use AI to _____?", start by asking "How might we solve _____?"[16] and whether using AI is the best way to solve the problem. After mapping your jobs and tasks into a user journey, you will need to determine which tasks require AI, which are meaningfully enhanced by AI, and which tasks do not benefit from AI or are even degraded by it.

When Does It Not Make Sense to Use AI

Not all problems need AI. Many products feel smart or intelligent without using AI. Sometimes a simpler solution that does not need AI can work just as well as if not better than an AI-based solution. The simpler solution has an added benefit of being easier to build, explain, maintain, and debug.[17] Here are some cases when it does not make sense to use AI (yet):

1. Maintaining predictability

2. Minimizing costly errors

3. Complete transparency

---

[16] pair.withgoogle.com, `https://pair.withgoogle.com/`.

[17] pair.withgoogle.com, `https://pair.withgoogle.com/`.

4. Optimizing for high speed

5. Optimizing for low costs

6. Static or limited information

7. Data being sparse

8. Social intelligence

9. People not wanting AI

Maintaining Predictability

Sometimes being predictable is a core part of a user experience. With some products, users form habits when using an interface or a workflow. For example, you can find the search button or a logout option in a menu in predictable places; their position rarely changes. Using AI to change the expected experience can break user habits that leads to frustration and causes cognitive load.

Minimizing Costly Errors

It would be better not to use AI if the cost of any errors outweighs the benefits that the AI provides. For example, the cost of misdiagnosing cancer outweighs any incremental improvement in cancer detection. It would lead to uncertainty in decision-making and cause users to trust the system less.

Complete Transparency

AI often operates in a black box and can't always provide a precise level of transparency. Using AI is not advisable if your team or customers need full transparency and explanations for how the AI system works or makes predictions, like criminal sentencing or open source projects.

Optimizing for High Speed

Building any robust AI system requires time for development and testing. Building an AI system from scratch would not be ideal if your goal is to get to the market first.

Optimizing for Low Costs

There is a tradeoff between the costs and the quality of your AI product. Building robust AI systems from scratch generally requires lots of data, computing power, and smart people; this can be expensive, especially if you are optimizing for lower costs.

Static or Limited Information

It would not be helpful to build AI when providing static or limited information. For example, visa application forms don't change for a large set of users.

Data Being Sparse

In most cases, the quality of your AI model is directly proportional to the amount of data available. It might not be suitable to use AI if you don't have adequate data to train your model. It is difficult for AI to learn complex concepts from small amounts of data or perform in environments it was not trained for.[18]

Social Intelligence

According to Moravec's paradox, easy things for a five-year-old are hard for machines and vice versa. Beating humans at chess is much easier for an AI than having a long conversation. It is not ideal to use AI for situations

---

[18] Ng, Andrew. AI for Everyone. www.coursera.org/learn/ai-for-everyone.

that require social intelligence, like interacting with other people. People can be sarcastic; they can change their tone of voice to convey conflicting meanings. They use non-verbal signals like a wink or frown to express emotions. AI is not great at understanding these subtle cues that people use when interacting with each other.

People Not Wanting AI

People like to do certain tasks, especially enjoyable ones like painting, reading, hiking, etc. If people explicitly tell you that they don't want to use AI, it might not be advisable to disrupt that activity, even if AI can handle it.

When Does It Make Sense to Use AI

There are many scenarios where using AI can be extremely beneficial. AI can significantly improve the user experience, speed, and quality of outcomes. It can help scale tasks that can take a lot of time to complete manually. According to Andrew Ng, AI would be good at solving problems that take humans less than a second to complete and where lots of data are available. Something that takes less than one second of thought can be automated now or soon.[19] Although this is not a hard-and-fast rule, it serves as a good starting point for identifying tasks where using AI makes sense. The following are some cases where using AI would be beneficial:

1. Personalization

2. Recommendation

3. Recognition

4. Categorization and classification

5. Prediction

---

[19] Ng, Andrew. AI for Everyone. www.coursera.org/learn/ai-for-everyone.

6. Ranking

7. Detecting anomalies

8. Natural language understanding (NLU)

9. Generating new data

Personalization

It is good to use AI when personalizing the experience for different users. For example, you can use AI to recommend personalized content like in the case of Netflix or change the settings of a car like the temperature, angle of steering, height of seats, etc., based on who is inside.

Recommendation

You can use AI to recommend different content to different users. For example, Spotify suggests what songs to listen to next, or Amazon recommends which books to buy based on previous purchases and similar buying patterns. You can also use recommendations to surface content that would otherwise be impossible for users to find on their own. Many social media applications like TikTok or Instagram use recommendation algorithms to generate a continuous stream of dynamic content.

Recognition

AI systems can recognize a class of entities from lots of data like cats from videos or faces of people from photos. Many products like Facebook use AI recognition systems for tagging photos; even self-driving cars use AI to recognize vehicles and people on the road.

Categorization and Classification

You can use AI to categorize entities into different sets. Categorization can be used for many applications like sorting vegetables from a pile or detecting faulty products in an assembly line or by photo apps to classify images into landscapes, selfies, etc. Clustering is a common technique used for generating a set of categories. For example, in an advertising company, clustering can be used to segment customers based on demographics, preferences, and buying behavior.

Prediction

You can use AI to predict future events or the next course of action. Prediction is the process of filling in the missing information. Prediction takes the information you have, often called "data," and uses it to generate information you don't have.[20] For example, AI models are used in systems that predict flight prices or predict the number of orders while managing inventory. A ubiquitous example is the smartphone keyboard that uses predictive models to suggest the next word.

Ranking

You can use AI to rank items, especially when it is difficult to determine a clear ranking logic. Ranking algorithms are used in search engines to decide the order of results. PageRank is an algorithm used by Google Search to rank web pages in search results. Ranking is also used along with other AI applications like product recommendation systems to decide the order of items suggested.

---

[20] Agrawal, Ajay, et al. *Prediction Machines*. Harvard Business Review Press, 2018.

Detecting Anomalies

AI systems are good for detecting anomalies in large amounts of data. Detecting anomalies means determining specific inputs that are out of the ordinary. For example, AI systems can help radiologists detect lung cancer by looking at X-ray scans. Banks use software that detects anomalous spending patterns to detect credit card fraud.

Natural Language Understanding

You can use AI for various natural language tasks like translation or converting speech to text. Smart assistants like Amazon's Alexa or Apple's Siri use natural language understanding to respond to queries. Google Translate uses NLU to translate between languages.

Generating New Data

You can sometimes use AI to generate new information based on training data. You can use techniques like generative adversarial networks (GANs) to generate graphics or music in a particular style after being trained on data from a similar style.

These are some of the most common types of tasks where using AI is beneficial. Often your teams would use multiple techniques to achieve the desired outcome. For example, any prediction or recommendation system would produce a list of results that will need to be ordered based on a ranking algorithm.

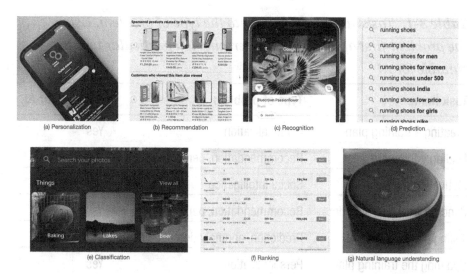

Figure 3-8. Examples of AI use cases. *(a) Personalized AI-generated playlists on Spotify. Source: Photo by David Švihovec on Unsplash. (b) Amazon product recommendations. (c) Google Lens recognizing the type of plant from an image. Source:* `https://play.google.com/store/apps/details?id=com.google.ar.lens` *(d) Google Search automatically predicts the query. (e) Google Photos automatically sorts images by things in them. (f) Flight booking website ranks results by relevance. (g) Amazon Alexa is responding to a voice command. Source: Photo by Lazar Gugleta on Unsplash*

Identifying Tasks Suitable for AI

Once you've identified jobs and tasks and mapped those onto a user journey, the next step is to determine which tasks can be handled meaningfully by AI. It would be ideal to use the problem-first approach discussed previously to find suitable tasks.

Let's take the previous example of a person training for a marathon. Using the problem-first approach, we can identify tasks and select which are suitable for AI. Refer to the following table.

Table 3-1. *Tasks are categorized into the type of problem. Using the problem-first approach, we determine if a particular task is suitable for AI*

Task name	Type of problem	Suitable for AI
Creating a training plan	Personalization	Yes
Finding equipment	Recommendation, personalization	Yes
Motivating the trainee	Social intelligence	No
Sending workout reminders	Recommendation, personalization	Yes
Telling progress	Prediction, anomaly detection	Yes
Updating the training plan	Personalization	Yes

Considerations for AI Tasks

Figuring out tasks that are suitable for AI is a critical part of identifying AI opportunities. These are all possible opportunities, but they might not all be feasible. Once you've found out tasks where it makes sense to use AI, you need to figure out the level of automation required and whether it is practical to build AI capabilities. The following are a few important considerations that you need to think of before designing any AI solution:

1. Type of action

2. Type of environment

3. Data

4. Cost

5. Time and effort

6. Quality improvements and gains

7. Societal norms

Type of Action

Once you've found tasks suitable for AI, the next step is to evaluate whether AI should automate the task or augment the user's ability to perform the task.

- *Augmentation*: AI helps people in completing a task.

- *Automation*: AI completes the task on behalf of people.

Although it can be tempting to say that AI will automate everything, there are many activities that people will want to do themselves. In such cases, AI can help people perform the activity faster and more efficiently or creatively.

Augmentation

With augmentation, the AI system extends the ability of people to perform a task. It is good to augment tasks that are inherently human, personally valuable, or high stakes in nature. Think of giving your users superpowers instead of doing the work for them.[21]

Here are examples of augmentation:

1. An illustration design software that suggests colors to pick based on what's on the artboard.

2. Drones have become a fascinating application of AI: computer vision and intelligent algorithms process video in real time, thereby allowing people to extend their vision and delivery capabilities up into the air and over miles of potentially impassable terrain.[22]

---

[21] pair.withgoogle.com, `https://pair.withgoogle.com/`.

[22] Wilson, H. James, and Paul R. Daugherty. *Human + Machine: Reimagining Work in the Age of AI*. Harvard Business Review, 2018.

3. On the assembly line, workers can collaborate with a self-aware robot arm. Industrial work becomes decidedly less manual and more like being a pilot, where the robot becomes an extension of the worker's body. In a study with Mercedes's competitor BMW, researchers determined that human-robot interactions in the car plant were about 85% more productive than either humans or robots on their own.[23]

4. A designer at Autodesk who decides to build a drone. Instead of modifying preexisting concepts and adjusting for various constraints like weight and propulsion, she inputs these parameters into the company's AI-enabled software. The software's genetic algorithm produces a vast and dizzying array of new designs that no one has ever seen. Some are more bizarre than others, but all fit the initial constraints.[24]

5. A tool like Grammarly suggests writing tips as you're writing an email or a document.

When to Augment

It isn't always easy to separate augmentation from automation. However, augmentation opportunities are usually more complicated, inherently human, and personally valuable.[25] Here's when it makes sense to augment:

---

[23] Wilson, H. James, and Paul R. Daugherty. *Human + Machine: Reimagining Work in the Age of AI.* Harvard Business Review, 2018.

[24] Wilson, H. James, and Paul R. Daugherty. *Human + Machine: Reimagining Work in the Age of AI.* Harvard Business Review, 2018.

[25] pair.withgoogle.com, https://pair.withgoogle.com/.

1. **People enjoy the task.**

    Not every task is a chore. If you enjoy writing poetry, you wouldn't want an AI to write it for you. You wouldn't like it if an algorithm took over the creative process. However, AI can help by recommending words and phrases and help structure sentences in such cases.

2. **Stakes of the situation are high.**

    When the stakes are high, people prefer to stay in control. People need human control and oversight in essential or critical situations like performing medical surgery or managing personal finances. You wouldn't trust an AI with even a 1% chance of performing surgery incorrectly, leading to death, or a 2% chance of losing all your life savings. In such cases, AI can help the surgeon by detecting anomalies or a portfolio manager by recommending investment opportunities.

3. **Personal responsibility is needed.**

    Sometimes a personal responsibility for an outcome is needed or required. You wouldn't want an AI to give out harsh feedback to an employee or family member. Sometimes a human responsibility is legally required, like in the case of policing or driving. In such cases, an AI can recommend an action, but taking it requires human judgment.

4. **Specifics are** hard **to communicate.**

   Sometimes people can imagine how a system
   should do something, but it is hard to communicate
   specifics. Tastes are subjective. In such cases,
   people prefer to stay in control. For example, when
   designing home interiors or making art, it is difficult
   to specify the vision of how you want it. In such
   cases, AI can help by recommending options of
   furniture, colors, etc.

Measuring Successful Augmentation

Augmentation is successful when

1. **People enjoy doing the task more.** There is a
   greater sense of responsibility and fulfillment. You
   can measure enjoyment through feedback surveys,
   NPSs, or conducting user interviews.

2. **The creativity of users has increased.** Creativity
   is difficult to measure quantitatively, but you can
   infer this by conducting contextual inquiries, user
   interviews, and customer feedback.

3. **Users can scale their efforts.** You can measure this
   as an improvement in the speed of performing a
   task or increased output.

4. **Users have higher levels of control over
   automated systems.** You can measure this by a
   decrease in the number of errors or mistakes.

Automation

With automation, AI performs a task without user involvement. It is good to automate undesirable tasks where the user's investment of time, money, or effort is not worth it. Many of the tasks that can be automated do not require oversight, and users are often happy to delegate.[26]

Here are examples of automation:

1. A photo library application automatically sorts photos by people.

2. Spotify recommends what songs to listen to or automatically creates playlists for you.

3. Finding an Uber ride by specifying pickup and destination.

4. An email application automatically categorizing spam.

5. Recommending content on social media applications like Instagram or TikTok in the form of reels or stories.

6. Amazon suggests products to buy based on past purchases.

When to Automate

Automation is preferred when people want to delegate undesirable tasks or when the investment of time, effort, or money isn't worth it. People are happy to hand over such tasks that don't require human oversight and can be done just as well by someone or something else. Automation is

---

[26] pair.withgoogle.com, https://pair.withgoogle.com/.

often the best option for tasks that supplement human weaknesses with AI strengths.[27] Here's when it makes sense to automate:

1. **People lack the knowledge or ability to do a task.**

   There are many times when people don't know how to do a task. Sometimes they know how to do it, but a machine would do a better job, like sorting images of people from thousands of photos.

2. **There are temporary limitations.**

   There are cases when people don't have the time to do a task and don't mind delegating. In such cases, people would prefer to give up control. for example, when they are booking a ride to the airport while getting ready.

3. **Stakes of the situation are low.**

   In low-stakes situations like getting song or movie recommendations from a streaming service, people tend to give up control because the prospect of discovery is more important than the low cost of error.[28]

4. **Tasks are boring.**

   People prefer to delegate tasks that they find boring. For example, it would be incredibly boring to sort thousands of photos into categories. Even tasks like tax calculations can be tiresome.

---

[27] pair.withgoogle.com, https://pair.withgoogle.com/.

[28] pair.withgoogle.com, https://pair.withgoogle.com/.

5. **Tasks are repetitive.**

   People and organizations prefer to automate repetitive tasks. For example, transcribing audio samples is a repetitive task that you can hand off to an AI.

6. **Tasks are awkward.**

   People like to delegate awkward or undesirable tasks like asking for money or calling customer care.

7. **Tasks are dangerous.**

   People prefer machines to do tasks that are dangerous for people. It's unwise to check for a gas leak in a building using your own nose when you could use a sensor to detect the leak.[29]

Measuring Successful Automation

Automation is successful when

1. **Efficiency is increased.** You can measure efficiency through an increase in the speed of completion or output of tasks.

2. **Human safety has improved.** You can measure safety by a decrease in accidents in dangerous environments.

---

[29] pair.withgoogle.com, https://pair.withgoogle.com/.

3. **Tedious tasks are reduced.** You can measure this by the improvement in the output of tasks. Reduction in tedious tasks can lead to greater satisfaction, which you can gauge through NPSs, feedback surveys, or customer interviews.

4. **New experiences are enabled.** Sometimes automation can enable new experiences. Innovation is difficult to measure quantitatively, but you can infer this by conducting a contextual inquiry, user interviews, and customer feedback.

Human in the Loop

AI systems are probabilistic and can make mistakes. Automation will not always be foolproof. Even when you choose to automate a task, there should almost always be an option for human oversight—sometimes called "human in the loop"—and intervention if necessary. Easy options for this are allowing users to preview, test, edit, or undo any functions that your AI automates.[30] When your system isn't certain or can't complete a request, make sure there's a default user experience that doesn't rely on AI.[31]

Example: Training for a Marathon

Continuing our previous example of a person training for a marathon, let's look at different tasks that can be augmented or automated.

---

[30] pair.withgoogle.com, `https://pair.withgoogle.com/`.

[31] Kore, Akshay. "Designing AI products." woktossrobot.com, 2020, `https://woktossrobot.com/aiguide/`.

Table 3-2. *We mention the characteristics of a task and determine if the task needs to be automated or augmented*

Task name	Suitable for AI	Characteristics of task	Type of action
Creating a training plan	Yes	Boring, lack of knowledge, tedious	Automate
Finding equipment	Yes	Lack of knowledge; specifics are hard to communicate	Automate and/or augment
Motivating the trainee	No	Not applicable	Not applicable
Sending workout reminders	Yes	Repetitive	Automate
Telling progress	Yes	Repetitive; specifics may be hard to communicate	Automate and/or augment
Updating the training plan	Yes	Specifics may be hard to communicate	Augment or don't use AI

Type of Environment

The type of environment in which the AI operates can influence your design decisions. There are different lenses through which you can categorize environments. You can use this categorization to determine design decisions.

Full or Partial Observability

Environments may be fully observable or partially observable. Most computer or board games like chess are fully observable, while most real-world situations like driving a car or running an organization are partially observable. It is generally easier to build AI systems for fully observable environments where you can automate a large number of tasks. In partially observable environments, most tasks need human oversight or augmentation.

Continuous or Discrete Actions

Actions in an environment may be continuous, like driving, or discrete, in the case of chess. Continuous environments are harder to design for since the nature of the problem keeps changing. For example, driving a car presents different issues depending on the traffic, number of pedestrians, weather, road conditions, etc. AI in continuous environments requires human oversight, and most AI actions would be augmented. In the case of discrete actions, you can largely automate AI tasks.

Number of Agents

Whether the environment contains a single agent or multiple agents would determine the number of stakeholders you need to consider when designing a solution. Driving is a multi-agent environment, while finding the shortest route on a map involves only a single agent.

Predictable or Unpredictable Environments

In predictable environments, outcomes of actions are known or can be determined by the rules of the environment. Chess is an example of a predictable environment. For unpredictable environments, outcomes are uncertain and cannot be determined by the rules of the environment. Traffic or weather predictions are examples of unpredictable environments. It is generally easier to build AI systems for predictable environments. You can automate many AI tasks in predictable environments, while most AI tasks in unpredictable environments require human oversight and can be augmented.

Dynamic or Static Environments

Environments may be dynamic or static. When the environment is dynamically changing, like traffic or weather, the time to make a decision is limited, like driving. For static environments, there is enough time available to make decisions like performing complex tax calculations. It is generally harder to build AI systems for dynamic environments. AI tasks in dynamic environments may be automated when users don't have time or augmented if the stakes are high. For static environments, you can automate many of the AI tasks.

Time Horizon

Time horizon is the length of time within which an AI needs to make decisions. It can be short duration like emergency braking, intermediate duration in the case of chess, or long duration when training for a marathon. There are only a few tasks for short-duration horizons. With intermediate- or long-duration-time horizons, the number and variety of tasks would be higher.

Data

All current AI systems depend on data, and a large number of them need data that is labeled. When you look at the places where machine learning has made a difference, it's really been where we have an accumulation of large amounts of data, and we have people who can think simultaneously about the problem domain and how machine learning can solve that.[32] When thinking of using AI for tasks, you need to ask two questions about data:

1. Is the data available?

2. Can I access it?

---

[32] Ford, Martin R. *Architects of Intelligence*. Packt, 2018.

Availability

When building AI solutions, along with the algorithm, data is a critical component. You cannot build AI solutions if no data is available. Then you need large quantities of it to get any meaningful results. Many of these AI techniques still largely rely on labeled data, and there are still lots of limitations in terms of the availability of labeled data.[33] Sometimes when no data is available, teams can choose to build a dataset by manually or algorithmically generating data.

1. For manual data labeling, organizations can choose to employ data labeling services. Most companies working on this technology employ hundreds or even thousands of people, often in offshore outsourcing centers in India or China, whose job it is to teach the robo-cars to recognize pedestrians, cyclists, and other obstacles. The workers do this by manually marking up or "labeling" thousands of hours of video footage, often frame by frame.[34]

2. Sometimes, data can also be generated algorithmically, that is, by the AI itself. If you can simulate the world you're working in, then you can create your own training data, and that's what DeepMind did by having it play itself.[35] This type of a data generation system works well for games where the environments are fully observable, discrete, static, and predictable.

---

[33] Ford, Martin R. *Architects of Intelligence*. Packt, 2018.

[34] Mitchell, Melanie. *Artificial Intelligence*. First ed., Farrar, Straus and Giroux, 2019.

[35] Ford, Martin R. *Architects of Intelligence*. Packt, 2018.

Access

Even if the data is available, you need a way to access it. For many applications, the data might be present in another company's database. Data is a strategic asset, and companies might not want to share it. Sometimes there are good reasons for the data to not be available, like privacy, regulations, or security. Generally, there are two ways to access data.

Access from Within

This is the case when data is available within your company or you can generate it as we discussed earlier. If AI is at the center of your product strategy, then you need to control the data to improve the system. In such cases, you can consider building a dataset.

External Access

Organizations can often access data from external sources like a public or private database. They can either buy a product or access an API. Accessing data externally works well if AI is not at the center of your strategy. You can treat this data like electricity and purchase it off the shelf from the market.

Compounding Improvements

Data is an asset that creates a defensible barrier for your product. It is really difficult to recreate Google Search because you would need the amount of data Google already has. AI improvements are compounded as you add more data to the system. As one Google Translate engineer put it, "When you go from 10,000 training examples to 10 billion training examples, it all starts to work. Data trumps everything."[36]

---

[36] Kasparov, Garry, and Mig Greengard. *Deep Thinking*. John Murray, 2018.

At the same time, getting a machine system to a 90% effectiveness rate may be enough to make it useful, but it's often even harder to get it from 90% to 95%, let alone to the 99.99% you would want before trusting it to translate a love letter or drive your kids to school.[37] Data has decreasing returns to scale. You get more useful information from a third of observation than a hundredth. Beyond a threshold, adding more units to your training data often becomes less useful in improving your prediction.[38]

Cost

The monetary cost of building AI is an important consideration. It will not make sense to build AI if the cost of building and maintaining AI is greater than the value of augmenting or automating a task. Cost can impact whether it is feasible to use AI. There are three critical costs associated with building AI:

1. **Cost of data**

   Data collection is costly and is often a significant investment. The cost depends on the amount of data needed and the detail in which you need to capture it.

2. **Human resources**

   You need skilled people to build and maintain AI. You can hire them from outside the company, or you could train your current employees. Both approaches involve monetary costs.

---

[37] Kasparov, Garry, and Mig Greengard. *Deep Thinking*. John Murray, 2018.
[38] Agrawal, Ajay, et al. *Prediction Machines*. Harvard Business Review Press, 2018.

3.   **Machine resources**

> You also need hardware and compute power to
> deploy AI solutions, which incur costs.

The costs of building AI can also vary depending on the nature of the task. Typically, when you're replacing cognitive work, it's mostly software and a standard computing platform.[39] The marginal cost economics can come down quickly and not cost too much. However, when replacing physical work, you might need to build a physical machine with hardware. While the cost of it might come down, it would not come down as fast as a software-only solution. It is technically easier to automate large portions of what the accountant does, mostly data analysis, data gathering, and so forth, whereas it's still technically harder to automate what a gardener does, which is mostly physical work in a highly unstructured environment.[40] Because society pays information workers like accountants more than physical workers like gardeners, the incentive to automate the accountant is already higher. We've started to realize that many of the low-wage jobs might be harder to automate from a technical and economic perspective.

Time and Effort

You need lots of time and effort to train, test, and deploy any meaningful AI solution. Additionally, you need to put in the time and effort to design a solution and collect data. If you don't already have the capability in your team, you'll need to put in time and effort to hire from outside or train your team. You need to consider if building an AI solution is the best use

---

[39] Ford, Martin R. *Architects of Intelligence*. Packt, 2018.
[40] Ford, Martin R. *Architects of Intelligence*. Packt, 2018.

of the time and effort of your team. Sometimes your problem may not be significant enough for the organization, and sometimes a simpler, non-AI solution can work just as well.

Quality Improvements and Gains

AI can lead to significant improvements in the quality of outcomes. These types of gains are beyond cost, time, and effort considerations. You can consider using AI for some tasks not because you are trying to reduce costs or labor but you're getting significantly better results or even superhuman outcomes. These are cases when you're getting better predictions or workflow improvements that you couldn't get with human capabilities. When you start to go beyond human capabilities and see performance improvements, that can really speed up the business case for AI deployment and adoption.[41]

Societal Norms

Societal norm is a broad term we can use for social acceptance factors and regulation in different cultures, societies, or nations. For example, in Japan, it is much more acceptable to have robots or machines in social settings than in other countries. Most of us wouldn't be comfortable with a medical diagnosis from a robot. Local regulations can also constrain AI use, which can differ by country or sector. For example, use and storage

---

[41] Ford, Martin R. *Architects of Intelligence*. Packt, 2018.

of personal information is especially sensitive in sectors such as banking, healthcare, and pharmaceutical and medical products, as well as in the public and social sectors.[42] You need to consider the norms of the society in which your AI product will be used. Different cultures will have different acceptance levels for AI, which can impact decisions like the level of automation and augmentation or even whether to build the AI solution.

Note   When finding tasks feasible for AI, the preceding steps and considerations are merely guidelines, not rules. A big part of being an AI product person is exercising judgment when to follow the rules and when to break them.

---

[42] Chui, Michael, et al. www.mckinsey.com/, 2018, www.mckinsey.com/featured-insights/artificial-intelligence/notes-from-the-ai-frontier-applications-and-value-of-deep-learning.

Here's a flowchart outlining the steps we've discussed so far.

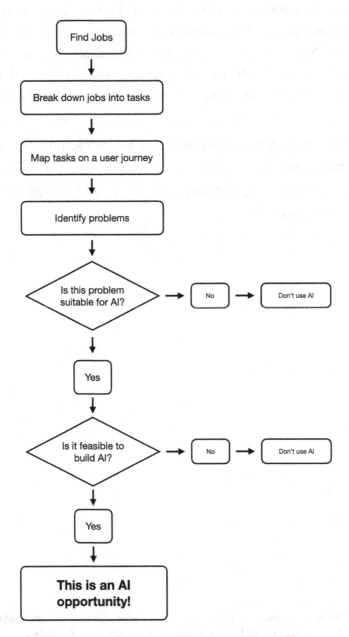

Figure 3-9. *Flowchart—how to find AI opportunities*

Big Red Button

AI is not perfect. When we say that an AI is 95% accurate, five out of a hundred results will be incorrect. You can trust such a system for low-stakes situations like movie recommendations, but you would be wary of trusting it for conducting surgery or diagnosing cancer. All AI systems are probabilistic and therefore error-prone, and they are bound to make mistakes at some point.

> *We need ways for people to oversee the system and take control if things go wrong.*

While an AI system is excellent at churning data, finding patterns, or detecting anomalies, it has no understanding of what it is doing. Humans are better at handling uncertain and unpredictable situations. Take the example of a judge deciding a person's fate—sentence them to jail or free them. Criminal judgment is often a difficult, moral decision. Even if an AI judge is 99% accurate, this means that one person out of a hundred is incorrectly sentenced, which in itself is wrong. Most of us would still prefer an imperfect human being as a judge rather than an AI that doesn't understand what it is doing.

Despite these shortcomings, AI has its advantages in improving the speed and quality of outcomes. Sometimes these improvements are superhuman. It would make sense to use AI in many cases while accounting for a margin of error. To account for errors, we need ways for people to oversee the system and take control if things go wrong. We need humans in the loop, that is, a big red button.

Levels of Autonomy

We looked at the concepts of augmentation and automation earlier. In augmentation, the AI helps users complete a task, while for automation, the AI completes the task on behalf of the user. In both these situations, human judgment is critical, and we need to account for a margin of error. To do that, we need the ability for humans to oversee, step in, or take over if things go wrong.

One of the goals of AI as a field is to build fully autonomous systems. However, we are yet to achieve complete automation in any of the current AI systems, and for a good reason. More AI autonomy means less control and the possibility of oversight.[43] For people to give up control over their decision-making, they would need to trust the AI more. Rather than thinking of full automation as binary, that is, systems are autonomous or not, a better mental model is to think in levels of autonomy. The higher the level, the more autonomous the system. Different organizations use a different number of steps. With self-driving cars, researchers generally work on six levels of autonomy from level 0 to level 5. For our purposes, we will use the ten levels[44] going from no AI to a fully autonomous solution.

1. *Level 1*: The computer offers no assistance, and humans make all decisions and actions. There is no AI used in this step.

2. *Level 2*: The AI offers a complete set of alternatives for decisions/actions. You can see this with Google Search, where the AI provides a choice of results.

3. *Level 3*: The AI narrows down the selection to a few alternatives. Netflix does this by showing you a personalized set of relevant movies.

4. *Level 4*: The AI suggests one decision, for example, Grammarly suggesting how to structure a sentence.

---

[43] Agrawal, Ajay, et al. *Prediction Machines*. Harvard Business Review Press, 2018.
[44] Kaushik, Meena, et al. UXAI. uxai.design, 2020, www.uxai.design/.

5. *Level 5*: The AI executes the suggestion if approved by the human, for example, asking Alexa to set an alarm.

6. *Level 6*: The AI allows the human a restricted time to veto before the automatic decision.

7. *Level 7*: The AI executes automatically and informs the human, for example, robots in warehouses.

8. *Level 8*: The AI informs the human only if asked, for example, spam filters.

9. *Level 9*: The AI informs the human only if the system decides to, for example, credit card fraud detection.

10. *Level 10*: The AI decides everything and acts while ignoring the human. Level 10 is full autonomy. We don't yet have any systems that operate without any human intervention.

Levels 2–5 are cases for augmentation, while levels 6–10 are cases for automation. You need a human in the loop at all levels except level 10. The level of autonomy is directly proportional to the amount of trust people place in the AI system.

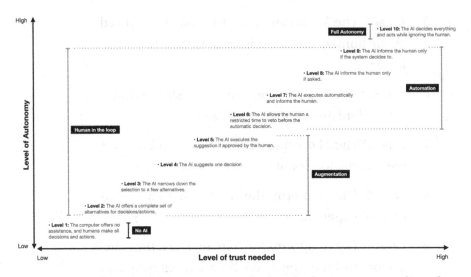

Figure 3-10. Levels of autonomy. *The level of autonomy is directly proportional to the amount of trust people place in the AI system. For level 10 or full autonomy, you need complete trust in the AI. There is no AI used in level 1. Levels 2–5 are cases for augmentation, while levels 6–10 are cases for automation. You need a human in the loop at all levels except level 10*

Rethinking Processes

So far, we've seen that you can use AI to augment or automate tasks and jobs. Doctors are now using AI to diagnose and treat cancer. Electrical companies use AI to improve power-generating efficiency. Investors use it to manage financial risk. Oil companies use it to improve safety on deep-sea rigs. Law enforcement agencies use it to hunt terrorists. Scientists use it to make new discoveries in astronomy and physics and neuroscience.[45] Any good AI system improves over time to adapt to changing environments. This improvement happens when the AI learns through feedback from humans and the environment.

---

[45] Polson, Nicholas G., and James Scott. *AIQ*. Bantam Press, 2018.

Instead of thinking of AI as something that merely replaces people's tasks, people and AI can act as symbiotic partners in a system. AI can give humans superpowers, while people can help improve the AI. Although it can help improve existing processes, you can also think of using AI to rethink processes, workflows, and strategies completely. Let's understand this with two examples.

Netflix

Netflix is a video content streaming service with a library of movies and TV shows that you can choose to watch for a monthly fee. On the surface, it is not a radical idea. Many of us have used movie rentals in the past. What Netflix does better is not that it has an extensive library or produces its own content, but it makes it easy to find things to watch. What you like to watch will be different from what I like to watch. The core idea is that each person has different preferences and their movie library should be tailored accordingly. So your Netflix library experience should be different from my experience.

The company is also investing in AI to improve other processes related to content distribution, like generating personalized movie trailers.[46] This would help them create trailers faster and help improve operating costs.

Netflix uses AI techniques like recommendations and ranking to suggest personalized content. Creators of Netflix ensured a close relationship and feedback loop with its customers. The AI builds a symbiotic relationship with you by learning from your behavior on the platform. What you watch, when you watch it, how you rate content, how much do you watch, etc. can all be used as feedback for the AI to learn and improve. By doing this, you get the benefit of a personalized library and discovery platform. You teach the AI your preferences, and in turn, the AI suggests what to watch.

---

[46] Min, Sarah. "Coming Soon to Netflix: Movie Trailers Crafted by AI." cbsnews.com, 2019, www.cbsnews.com/news/netflix-trailers-made-by-ai-netflix-is-investing-in-automation-to-make-trailers/.

Mercedes

Many luxury car manufacturers allow you to customize their cars online and choose from an expansive array of features. This trend in customizable cars led to Mercedes rethinking their assembly lines. In the factory, robots are moving beyond rigid assembly lines toward the idea of organic teams that partner humans with advanced AI systems. This collaboration between workers and smart machines is leading to the reinvention of many traditional processes.[47] These teams of humans and robots can adapt on the fly to new data and market conditions.

For a long time, factory robots have been large pieces of machinery that perform specific dedicated tasks, usually kept away from people. Imagine a smaller, nimbler, and flexible robot working alongside people to perform generic tasks like picking up heavy objects, rotating them, etc. There is an emerging class of robots that can learn new tricks overnight without any explicit programming.[48] Factory workers can show it how to do a task, and with the help of cameras and machine learning, these robotic arms can figure out the most efficient way to pick up parts and pieces and set them down. To fulfill a customized order, employees can partner with robots to perform new tasks without manually overhauling processes or manufacturing steps. These new changes are baked into the system and are performed automatically. The AI gives people superpowers, while people help train it to perform new tasks.

AI can give humans superpowers, while people can help improve the AI.

---

[47] Wilson, H. James, and Paul R. Daugherty. *Human + Machine: Reimagining Work in the Age of AI*. Harvard Business Review, 2018.

[48] Knight, Will. "This Factory Robot Learns a New Job Overnight." MIT Technology Review, 2016, www.technologyreview.com/2016/03/18/161519/this-factory-robot-learns-a-new-job-overnight/.

You can design processes around people and AI as partners in a system. In both these examples, AI gives humans superpowers, while people help improve it. There is a potential for powerful collaborations between humans and AI. Those companies that are using machines merely to replace humans will eventually stall, whereas those that think of innovative ways for machines to augment humans will become the leaders of their industries.[49]

Summary

This chapter outlined methods and considerations when incorporating AI into your product or workflows. Here are some important points:

1. Computers and artificial intelligence enable a cognitive division of labor between humans and machines. Humans and machines can overcome their individual weaknesses and focus on their strengths. Machines are good at doing routine tasks quickly, sifting through large amounts of data to generate predictions or detect anomalies. At the same time, humans are better at interpreting complex information and making judgments with limited information.

2. A Supermind is a group of individuals acting together in ways that seem intelligent.[50] The "individual" in this case could be a person or a machine. Most large projects are done by groups of people and machines, often within organizations.

---

[49] Wilson, H. James, and Paul R. Daugherty. *Human + Machine: Reimagining Work in the Age of AI*. Harvard Business Review, 2018.
[50] Malone, Thomas W. *Superminds*. Oneworld, 2018.

3. An organization is a type of a Supermind whose
    goal is to achieve the best possible outcome—
    thereby acting intelligently. We call this collective
    intelligence—the result of a group of individuals
    acting together in ways that seem intelligent.[51]
    Collective intelligence is something that makes
    organizations effective, productive, adaptable, and
    resilient.[52]

4. Incorporating AI is a way of increasing the collective
    intelligence of organizations. There are four ways in
    which AI can increase the collective intelligence of
    an organization:

    a.   Improving machines

    b.   Automating redundant tasks

    c.   Improving machine-machine collaboration

    d.   Improving human-machine collaboration

5. When incorporating AI into your workflows, you
    can also think of AI as a part of the organization
    and assign it specific roles. You can give AI different
    levels of control, and these roles can range from
    tools, to assistants, and to AI managing a group
    of people.

6. You can break down the process of identifying
    opportunities for incorporating AI into a few steps:

---

[51] Malone, Thomas W. *Superminds*. Oneworld, 2018.

[52] Kore, Akshay. "Systems, Superminds and Collective Intelligence."
Wok Toss Robot, 2020, https://woktossrobot.com/2020/05/12/
systems-superminds-and-collective-intelligence/.

    a.   Breaking down jobs into tasks

    b.   Mapping user journeys

    c.   Checking if AI is suitable for the task

    d.   Checking the feasibility of using AI for the task

    e.   Introducing humans in the loop

7. The first step is to break down jobs into tasks. A job is a way of accomplishing a goal. A task is a unit of activity that an individual performs to do their job. A job is a collection of multiple tasks.

8. After identifying tasks, you need to map those tasks on a user journey.

9. An essential part of being an AI person is identifying problems that need and can be solved with AI. You need to start with the problem and determine whether using AI is the best way to solve it.

10. Here are some cases when it **does not make sense** to use AI:

    a.   Maintaining predictability

    b.   Minimizing costly errors

    c.   Complete transparency

    d.   Optimizing for high speed

    e.   Optimizing for low costs

    f.   Static or limited information

    g.   Data being sparse

    h.   Social intelligence

    i.   People not wanting AI

11. The following are some cases where it **makes sense** to use AI:

    a.   Personalization

    b.   Recommendation

    c.   Recognition

    d.   Categorization and classification

    e.   Prediction

    f.   Ranking

    g.   Detecting anomalies

    h.   Natural language understanding

    i.   Generating new data

12. Even if using AI is suitable for a task, it might not always be feasible. The following are a few critical considerations that you need to think of before designing any AI solution:

    a.   Consider if you should automate the task or augment people's capability in doing the task.

    b.   Determine the type of environment in which your AI will exist.

    c.   Check if training data is available and accessible.

    d.   Consider the cost of building the AI.

    e.   Determine the time and effort required to build the AI system.

   f.   Quality improvements and gains you get with using AI.

   g.   You need to consider the norms of the society in which your AI product will be used. Different cultures will have different levels of acceptance for AI.

13. All AI systems are probabilistic and are therefore error-prone. They are bound to make mistakes at some point. To account for errors, we need methods for people to oversee the system and take control if things go wrong. We need humans in the loop.

14. You can categorize AI systems by their levels of autonomy. In this chapter, we used a framework of ten levels from levels 1 to 10. The level of autonomy is directly proportional to the amount of trust people place in the AI system. For level 10 or full autonomy, you need complete trust in the AI. There is no AI used in level 1. Levels 2–5 are cases for augmentation, while levels 6–10 are cases for automation. You need a human in the loop at all levels except level 10.

15. Instead of thinking of AI as something that merely replaces people's tasks, people and AI can act as symbiotic partners in a system. AI can give humans superpowers, while people can help improve the AI. Although it can help improve existing processes, you can also think of using AI to completely rethink processes, workflows, and strategies.

PART 3

Design

CHAPTER 4

Building Trust

This chapter will discuss why building the right level of trust with users is critical when designing AI products. We will look at different strategies for building and maintaining trust and opportunities for trust-building in your user experience. We will also look at the advantages, disadvantages, and considerations for imbuing personality and emotion in your AI.

I'm a stickler about having chai and biscuits in the morning. Every week I go to the neighborhood grocery store to buy a regular packet of Britannia Marie Gold biscuits. When paying at the counter, I don't check prices; I simply hand out my card. The store owner keeps these biscuits stocked for me and sometimes even gives me a discount. Over the past few years, I have come to trust this little store.

Almost all successful human interactions are built on trust. You often spend more money on brands that you trust over cheaper alternatives, which is why companies spend a lot of money on building their brand. When you go to a doctor for a health checkup, the underlying assumption is that you can trust their diagnosis. Most of us prefer to deposit our savings with a trusted bank. We don't make large purchases on websites we don't trust. We tend to leave our kids to play in safe places with trustworthy people.

© Akshay Kore 2022
A. Kore, *Designing Human-Centric AI Experiences*,
https://doi.org/10.1007/978-1-4842-8088-1_4

Without trust, it would be debilitating to perform even the simplest of tasks. Without trust, almost all systems break, and society would fall apart. Trust is the willingness to take a risk based on the expectation of a benefit.[1] Trust in business is the expectation that the other party will behave with integrity. In a high-trust environment, people are honest and truthful in their communication. There is a fair exchange of benefits, and people operate in good faith. There is adequate transparency into the workings of the system. Insufficient transparency leads to distrust. Clear accountability is established; people don't play blame games. A successful team of people is built on trust, so is a team of people and AI.

> *A successful team of people is built on trust, so is a team of people and AI.*

Trust in AI

We can use AI systems in organizations to generate predictions, provide insights, suggest actions, and sometimes even make decisions. The output of AI systems can affect different types of stakeholders directly or indirectly. AI systems are not perfect. They are probabilistic and learn from past data. Sometimes they make mistakes, which is why we require humans to oversee them. For AI products to be successful, people who use them or are affected by them need to trust these systems.

Components of User Trust

When building trust with the users of your AI system, you are essentially trying to develop a good relationship. Your users need to have the right level of confidence when using or working alongside your AI. The following components contribute to user trust in your product:

---

[1] pair.withgoogle.com, https://pair.withgoogle.com/.

1. Competence

2. Reliability

3. Predictability

4. Benevolence

Competence

Competence is the ability of the product to get the job done. Does it improve the experience or address the user's needs satisfactorily? A good-looking product or one with many features that do not fulfill user needs is not competent. Strive for a product that provides meaningful value that is easy to recognize.[2] Google Search is an example of a competent product. It generally offers satisfactory results for your questions.

Reliability

Reliability indicates how consistently your product delivers on its abilities.[3] A reliable product provides a consistent, predictable experience that is communicated clearly. A product that performs exceptionally well one time and breaks down the next time is not reliable. Apple's iPhone is an example of a reliable product. It might not carry all the features its competitors have, but you can reasonably trust the ones it does have.

Predictability

A predictable interface is necessary, especially when the stakes are high. If the user comes to your product to perform critical, time-sensitive tasks, like quickly updating a spreadsheet before a client presentation, don't

---

[2] pair.withgoogle.com, https://pair.withgoogle.com/
[3] pair.withgoogle.com, https://pair.withgoogle.com.

include anything in your UI that puts habituation at risk.[4] A probabilistic AI-based solution that can break the user's habit is not ideal in such cases. However, suppose you think that users have open-ended goals like exploration. In that case, you can consider a dynamic AI-based solution that sometimes breaks user habits, for example, selecting a movie from dynamic AI-based suggestions.

Benevolence

Benevolence is the belief that the trusted party wants to do good for the user.[5] Be honest and upfront about the value your users and your product will get out of the relationship. Patagonia is a clothing brand that makes jackets that does a great job with benevolence. While their products can be expensive, they encourage people to reuse and repair their Patagonia clothes, giving a percentage of their sales for environmental causes. The company is upfront about its value to the customer.

Trust Calibration

AI can help people augment or automate their tasks. People and AI can work alongside each other as partners in an organization. To collaborate efficiently, your stakeholders need to have the right level of trust in your AI system.

---

[4] Fisher, Kristie, and Shannon May. "Predictably Smart—Library." Google Design, 2018, https://design.google/library/predictably-smart/.

[5] pair.withgoogle.com, https://pair.withgoogle.com/.

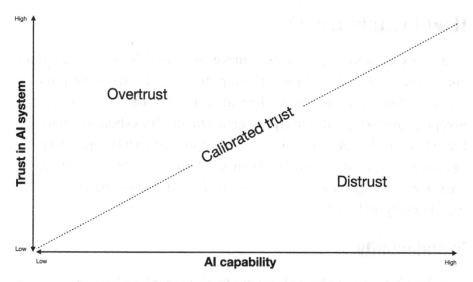

Figure 4-1. Trust calibration. *Users can overtrust the AI when their trust exceeds the system's capabilities. They can distrust the system if they are not confident of the AI's performance*

Users can sometimes overtrust or distrust your AI system, which results in a mismatch of expectations. Users may distrust your AI when trust is less than the system's capabilities. They may not be confident about your AI's recommendations and decide not to use them. Users rejecting its capabilities is a failure of the AI system. Overtrust happens when user trust exceeds the system's capabilities, leading to users trusting an AI's recommendation when they should be using their own judgment. For example, users overtrusting the suggestions of a stock prediction service can lead to a financial loss. Overtrust can quickly turn into distrust the next time this person uses the service.

Product teams need to calibrate user trust in the AI regularly. The process to earn user trust is slow, and it'll require proper calibration of the user's expectations and understanding of what the product can and can't do.[6]

---

[6]pair.withgoogle.com, https://pair.withgoogle.com/.

How to Build Trust?

AI systems are probabilistic and can make mistakes. It is the job of product creators to build trustworthy relationships between the AI and its users. Building trust is not about being right all the time; it is about integrity, accepting mistakes, and actively correcting them. Users should be able to judge how much they can trust your AI's outputs, when it is appropriate to defer to AI, and when they need to make their own judgments. There are two essential parts to building user trust for AI systems, namely, explainability and control.

Explainability

If we don't understand how AI systems work, we can't really trust them or predict the circumstances under which they will make errors.[7] Explainability means ensuring that users of your AI system understand how it works and how well it works. Your explanations allow product creators to set the right expectations and users to calibrate their trust in the AI's recommendations. While providing detailed explanations can be very complicated, we need to optimize our explanations for user understanding and clarity.

Control

Users should be able to second-guess the AI's predictions. Users will trust your AI more if they feel in control of their relationship with it. Giving users some control over the algorithm makes them more likely to feel the algorithm is superior and more likely to continue to use the AI system in the future.[8]

---

[7] Mitchell, Melanie. *Artificial Intelligence*. First ed., Farrar, Straus and Giroux, 2019.

[8] Wilson, H. James, and Paul R. Daugherty. *Human + Machine: Reimagining Work in the Age of AI*. Harvard Business Review, 2018.

You can do this by allowing users to edit data, choose the types of results, ignore recommendations, and correct mistakes through feedback.

Building trust is not about being right all the time; it is about integrity, accepting mistakes, and actively correcting them.

Explainability

We generally don't trust those who can't explain their decisions and reasoning. The same is true for AI systems that show recommendations, provide insights, or make decisions on behalf of people. Explaining their reason is especially important as AI systems significantly impact critical decisions like sanctioning loans, predicting diseases, or recommending jobs. When AI systems are used to help make decisions that impact people's lives, it is particularly important that people understand how those decisions were made.[9]

But even people can't always explain how they made certain decisions. You can't look "under the hood" into other people's brains. However, humans tend to trust that other humans have correctly mastered basic cognitive tasks. You trust other people when you believe that their thinking is like your own. In short, where other people are concerned, you have what psychologists call a theory of mind—a model of the other person's knowledge and goals in particular situations.[10] We don't yet have a theory of mind for AI systems.

When doing math homework, my teachers asked me to show the steps followed to reach an answer. Showing how I derived an answer allowed me to demonstrate my understanding. It helped my teachers know if I had learned the correct abstractions, arrived at a solution for the right reasons, and figured out why I made particular errors. Showing my work was a way

---

[9] Smith, Brad, and Harry Shum. *The Future Computed.* Microsoft Corporation, 2018.

[10] Mitchell, Melanie. *Artificial Intelligence.* First ed., Farrar, Straus and Giroux, 2019.

for me to present my decision-making process. Similarly, AI systems can benefit from showing how they arrived at a recommendation. An AI system providing a correct recommendation for the wrong reasons is a fluke; it is not trustworthy. Designing explanations can enable users to build the right level of trust in your AI system. Explainability and trust are inherently linked.

> *When AI systems are used to help make decisions that impact people's lives, it is particularly important that people under-stand how those decisions were made.*[11]

Who Needs an Explanation?

When designing AI products, you need to consider the different types of stakeholders in your system. Stakeholders can be the users of your system, those affected by the AI, and people responsible for monitoring the system. Different stakeholders in your AI system need different levels of explanations. There will likely be varying degrees of factors such as domain expertise, self-confidence, attitudes toward AI, and knowledge of how AI works, all of which can influence trust and how people understand the system.[12] By identifying and understanding your stakeholders, you can ensure that the explanation matches their needs and capabilities. There are four types of stakeholders in an AI system:[13]

1. Decision-makers

2. Affected users

3. Regulators

4. Internal stakeholders

---

[11] Smith, Brad, and Harry Shum. *The Future Computed.* Microsoft Corporation, 2018.

[12] Kaushik, Meena, et al. UXAI. uxai.design, 2020, www.uxai.design/.

[13] Kaushik, Meena, et al. UXAI. uxai.design, 2020, https://www.uxai.design/.

Decision-Makers

Decision-makers are people who use the AI system to make decisions. A decision-maker can be a bank officer deciding whether to sanction a loan or a radiologist diagnosing an ailment based on an AI's recommendation. Decision-makers need explanations to build trust and confidence in the AI's recommendations. They need to understand how the system works and sometimes require insights to improve their future decisions. Most decision-makers need simplified descriptions and have a low tolerance for complex explanations.

Affected Users

These are people impacted by recommendations that your AI system makes, like a loan applicant or a patient. Sometimes the decision-maker and the affected user are the same people. For example, giving Netflix your preferences is a decision, and getting movie recommendations is the effect. Affected users seek explanations that can help them understand if they were treated fairly and what factors could be changed to get a different result.[14] They have a low tolerance for complex explanations, and outcomes need to be communicated clearly and directly.

Regulators

These are people who check your AI system. Regulators can be within the organization in the form of internal auditing committees or external in the form of government agencies. Regulators can enforce policies like the European Union's General Data Protection Regulation (GDPR) and may require AI creators to provide explanations for decisions. Regulators need explanations that enable them to ensure that decisions are made in a fair and safe manner. They may sometimes use explanations to investigate

---

[14] Kaushik, Meena, et al. UXAI. uxai.design, 2020, www.uxai.design/.

a problem. Explanations for regulators can show the overall process, the training data used, and the level of confidence in the algorithm. These stakeholders may or may not have a high tolerance for complex explanations.

Internal Stakeholders

They are people who build the AI system, like ML engineers, product managers, designers, data scientists, and developers. Internal stakeholders need explanations to check if the system is working as expected, diagnose problems, and improve it using feedback. They generally have a high tolerance for complexity and need detailed explanations of the system's inner workings.

Explainability and trust are inherently linked.

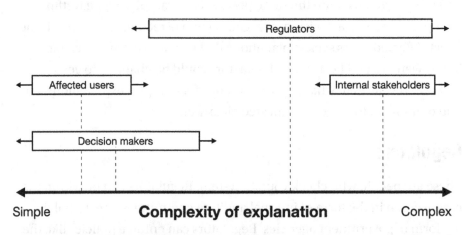

Figure 4-2. Tolerance for the complexity of explanations. Affected users and decision-makers often require more straightforward explanations, while regulators and internal stakeholders might not mind detailed or complex explanations

Guidelines for Designing AI Explanations

A mental model is a user's understanding of a system and how it works. Once users have clear mental models of the system's capabilities and limits, they can understand how and when to trust it to help accomplish their goals.[15] A mismatch between the user's understanding of how the system works and how it actually works can lead to frustration, misuse, and even product abandonment. Building the right mental models is key to user trust in your AI, and you can use explanations to calibrate it.

The process of designing AI explanations is the process of building the right mental models for your users. It is the process of explaining how the AI makes decisions and the relationship between their input and AI's output. Sometimes you can build effective mental models on top of existing ones. For example, the idea of a trash can is a good mental model for explaining where your files go when you delete them. The following are some guidelines that can help you design better AI explanations:

1. Make clear what the system can do.

2. Make clear how well the system does its job.

3. Set expectations for adaptation.

4. Plan for calibrating trust.

5. Be transparent.

6. Build cause-and-effect relationships.

7. Optimize for understanding.

---

[15] pair.withgoogle.com, https://pair.withgoogle.com/.

Make Clear What the System Can Do[16]

Make sure your users understand the capabilities of the system. Users need to be aware of the full breadth of functionality of your feature. Providing contextual information on how the AI system works and interacts with data can help build and reinforce user trust. For example, if there is a search functionality within a grocery application, make sure all search possibilities are enumerated, like "search for fruits, vegetables, and groceries." Clarify how input influences the results.

Try not to be opaque about how the system makes decisions, especially for high-stakes situations like loan sanctioning or predicting diseases. Unless your application can handle those, stay away from open-ended interaction patterns without proper guidance. For example, an AI running assistant asking questions like "Ask me anything?" or "What would you like to do today?" sets the wrong expectations about the system's capabilities.

---

[16] Kershaw, Nat, and C. J. Gronlund. "Introduction to Guidelines for Human-AI Interaction." Human-AI Interaction Guidelines, 2019, https://docs.microsoft.com/en-us/ai/guidelines-human-ai-interaction/.

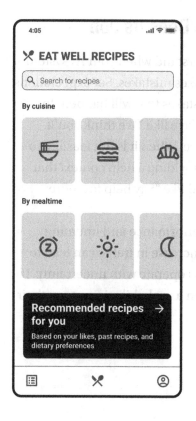

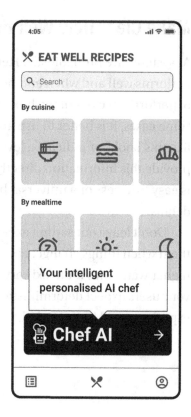

 Aim for    Avoid

Figure 4-3. Recipe recommendation app: Make clear what the system can do. (Left) Aim to set the right expectations. In this case, the interface clarifies that the AI has recommended personalized recipes based on some information. (Right) Avoid ambiguity in explanations. The term Chef AI is ambiguous and can set the wrong expectations of the AI in the context of a recipe recommendation app

Make Clear How Well the System Does Its Job

AI systems are probabilistic. Help users understand when your AI system performs well and when and how often it makes mistakes. Set expectations of performance out of the box and clarify mistakes that will happen. In some cases, it is better to use uncertain language like "We think you'll like this book" or indicate a level of confidence. When it is not feasible to provide this information directly, consider providing a help context that is easy to access or a universal help command like "Say help for more details."

Don't leave out setting expectations of performance and updating them when things change. For example, an increase in traffic can lead to higher wait times for a taxi service. AI systems operate with uncertainty. If your users expect deterministic behavior from a probabilistic system, their experience will be degraded.[17]

---

[17] Kershaw, Nat, and C. J. Gronlund. "Introduction to Guidelines for Human-AI Interaction." Human-AI Interaction Guidelines, 2019, https://docs.microsoft.com/en-us/ai/guidelines-human-ai-interaction/.

Aim for

Avoid

Figure 4-4. Plant recognizer: Make clear how well the system does its job. (Left) The interface explains the next action and sets expectations of performance. It tells the user what they can do and that the system works best for plants native to India. (Right) The tooltip on the right is ambiguous

Set Expectations for Adaptation

AI systems change over time based on feedback. Most AI systems adapt to changing environments to optimize their output or personalize to users. AI systems learn from their environments; sometimes, they keep track of whether or not an output was useful and adapt accordingly. For example,

Netflix adjusts its recommendations based on your interactions with the service like movies watched, for how long, user ratings, etc. Indicate whether the product adapts and learns over time. Clarify how input influences results. Don't forget to communicate changes in accuracy as conditions change. For example, a navigation system predicting the time to arrival can adjust depending on changes in traffic or weather conditions.

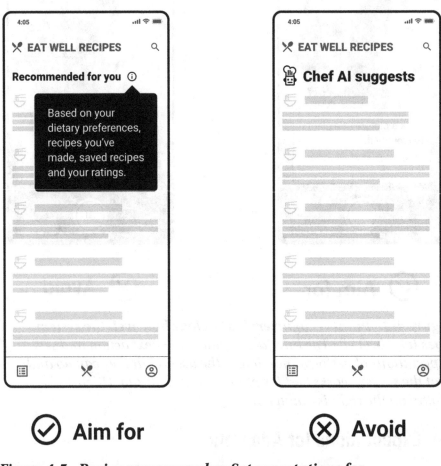

Figure 4-5.   Recipe recommender: Set expectations for adaptation. (Left) The tooltip explains how the system generated the recipe recommendations based on user preferences and activity. (Right) The recommendation on the right is ambiguous

Plan for Calibrating Trust

AI systems will not perform perfectly all the time. Users shouldn't implicitly trust your AI system in all circumstances but rather calibrate their trust correctly.[18] People may sometimes overtrust your system for something it can't do, or they may distrust the AI's output. Either of these cases causes a mismatch in expectations and can lead to the failure of the product. You can help users calibrate their trust in the AI by explaining what it can and can't do or displaying its output confidence. This uncertainty can lead to some level of distrust initially. For example, telling users that the prediction may be wrong can lead to them trusting the system less, but over time as the AI improves, they can start relying on your product more with the right level of trust.

The key to success in these human-AI partnerships is calibrating trust on a case-by-case basis, requiring the person to know when to trust the AI prediction and when to use their own judgment in order to improve decision outcomes in cases where the model is likely to perform poorly.[19] It takes time to calibrate trust with users. Plan for trust calibration throughout the user journey over a long time. AI changes and adapts over time, and so should the user's relationship with the product. For example, a music recommendation service may not be very accurate for new users. But over time, it improves by learning user preferences. Explaining that the service learns your preferences over time and will get better at recommendations can help users calibrate their trust at different stages of using the product. Planning for calibrating trust can help you in designing better explanations and build more effective human-in-loop workflows.

---

[18] pair.withgoogle.com, https://pair.withgoogle.com/.
[19] Kaushik, Meena, et al. UXAI. uxai.design, 2020, www.uxai.design/.

Here are some examples of workflows that help in calibrating trust:

1. Communicating what data is used to train the AI. A speech recognition system built on American speaker data might not perform well for Indian users.

2. Allowing users to specify language and genre preferences in a movie recommendation service.

3. Allowing users to try the product in a "sandbox" environment can help calibrate trust.

4. Displaying accuracy levels or a change in accuracy when recognizing product defects in an assembly line.

5. An ecommerce website showing reasons for product recommendations like "Customers who bought this also bought…" or "Similar products."

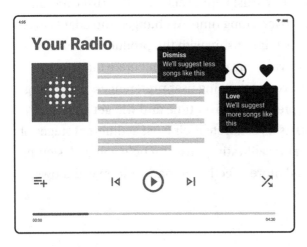

A personalised playlist based on your listening habits. Improves as you listen more.

Figure 4-6. Music recommendation. The messaging highlights how the AI improves over time

Plan for trust calibration throughout the user journey over a long time. AI changes and adapts over time, and so should the user's relationship with the product.

Be Transparent

It is difficult to trust those who appear to be hiding something. Transparency means operating in ways that make it easy for others to understand the actions of a system. It implies acting out in the open. There are legitimate reasons for organizations to have trade secrets or proprietary information. But when it comes to pertinent information of customers, privacy, data use, bias, other stakeholders, or the efficacy of the algorithm, transparency is central to earning trust. This is especially true for AI operating in high-stakes environments like medicine or driverless cars. A lack of transparency increases the risk and magnitude of harm when users do not understand the systems they are using or there is a failure to fix faults and improve systems following accidents.[20]

Many AI systems will have to comply with privacy laws that require transparency about the collection, use, and storage of data and mandate that consumers have appropriate controls so that they can choose how their data is used.[21] Product teams need to ensure that users are made aware of any data that is collected, tracked, or monitored and that it's easy for them to find out how the data is collected, whether via sensors, user-entered data, or other sources.[22] In necessary cases, use non–black box

---

[20] Ethically Aligned Design: A Vision for Prioritizing Human Well-Being with Autonomous and Intelligent Systems. First ed., IEEE, 2019.

[21] Smith, Brad, and Harry Shum. *The Future Computed.* Microsoft Corporation, 2018.

[22] Kershaw, Nat, and C. J. Gronlund. "Introduction to Guidelines for Human-AI Interaction." Human-AI Interaction Guidelines, 2019, https://docs.microsoft.com/en-us/ai/guidelines-human-ai-interaction/.

models so intermediate steps are interpretable and outcomes are clear, providing transparency to the process.[23]

The level of transparency will be different for different stakeholders. Your explanations can help users understand why the AI made certain predictions through transparency of data, its use, and the underlying algorithm. Good explanations can help people understand these systems better and establish the right level of trust.

> *Transparency means operating in ways that make it easy for others to understand the actions of a system.*

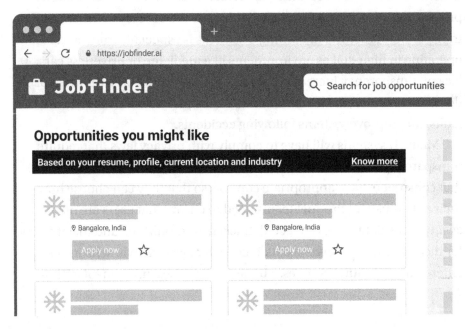

Figure 4-7. *Job recommendations. The callout explains how the system generated the recommendations*

---

[23] Wilson, H. James, and Paul R. Daugherty. *Human + Machine: Reimagining Work in the Age of AI.* Harvard Business Review, 2018.

Build Cause-and-Effect Relationships

People understand faster when they can identify a cause-and-effect relationship between their actions and the system's response. Sometimes the perfect time to show an explanation is in response to a user action. Users might be confused if an AI system completes an action but does not respond about its completion. If it reacts unexpectedly, an explanation can help in calibrating or recovering trust.

On the other hand, when the system is working well, responding to users' actions is a great time to tell the user what they can do to help the system continue to be reliable. For example, a user looks for personalized recommendations for breakfast places on a restaurant recommendation service like Zomato. If they only see recommendations for places they rarely visit or that don't match their earlier preferences, they might be disappointed and trust the service a bit less. However, suppose the app's recommendation includes an explanation that the system only recommends restaurants within a 2 km driving distance and that the user is standing in the heart of Cubbon Park in Bangalore. In that case, trust is likely to be maintained. The user can see how their actions affect the suggestion.

Building trust is a long-term process. A user's relationship with your product can evolve over time through back-and-forth interactions that reveal the AI's strengths, weaknesses, and behaviors.

Sometimes it can be hard to tie explanations to user actions, especially for interfaces like smart speakers. In such cases, you can use a multimodal approach to indicate understanding and response. For example, an Alexa smart speaker couples voice feedback with the light ring to indicate various responses and states. However, it is common for devices to have multiple modalities like visual, voice, etc., in the case of laptops, TVs, or smartphones. You can use a multimodal design approach even in such cases. For example, a smart TV with an assistant can respond to a query through voice but leave the explanation on the screen.

Building trust is a long-term process. A user's relationship with your product can evolve over time through back-and-forth interactions that reveal the AI's strengths, weaknesses, and behaviors.

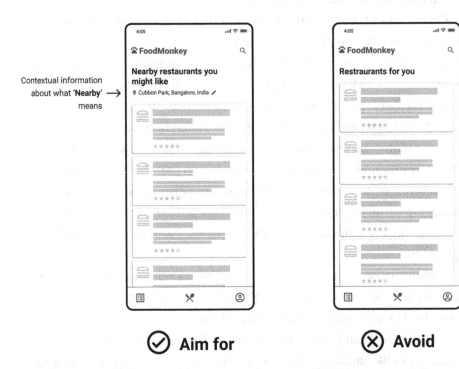

Contextual information about what **'Nearby'** means →

⊘ **Aim for**   ⊗ **Avoid**

Figure 4-8. Restaurant recommendations: Build cause-and-effect relationships. (Left) The restaurant recommendations include contextual information of the user's current location, which helps the user understand why certain restaurants are recommended. (Right) The recommendation on the right is ambiguous

Optimize for Understanding

Different stakeholders in your AI system need different levels of explanations. Your AI explanations need to be understandable. Simply publishing algorithms underlying AI systems or providing a dump of data used is not meaningful. A list of a billion operations is not an explanation that a human can understand.[24] The complexity of your explanations needs to be tailored for your users.

It can be challenging to explain how your AI system works. Providing a detailed explanation can sometimes confuse users. In such cases, the best approach is not to attempt to explain everything—just the aspects that impact user trust and decision-making.[25] Sometimes, it is required by law in certain regions like the EU for this information to be communicated "in a concise, transparent, intelligible and easily accessible form, using clear and plain language."[26]

The ability to explain can also determine the fate of your AI product. Product teams are responsible for making important judgment calls about which AI technologies might best be deployed for specific applications. A huge consideration here is accuracy vs. "explainability."[27] Deploying a deep learning system in high-stakes environments like medicine or driverless cars that we can't explain is risky. Such algorithms may require periodic regulatory scrutiny and sometimes might not be worth the cost of maintenance.

Explainability is critical for building user trust. Explaining your AI system so people can actually understand it is a fundamental human-centered AI design challenge.

---

[24] Mitchell, Melanie. *Artificial Intelligence*. First ed., Farrar, Straus and Giroux, 2019.
[25] pair.withgoogle.com, `https://pair.withgoogle.com/`.
[26] Mitchell, Melanie. *Artificial Intelligence*. First ed., Farrar, Straus and Giroux, 2019.
[27] Wilson, H. James, and Paul R. Daugherty. *Human + Machine: Reimagining Work in the Age of AI*. Harvard Business Review, 2018.

🚗 In-car navigation

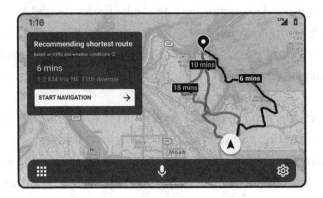

⊘ **Aim for**

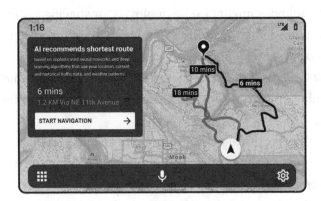

⊗ **Avoid**

Figure 4-9. *Don't explain everything. (Top) Informing the user that the shortest route is recommended is an easy explanation that most users can understand and act on. (Bottom) In this case, a detailed explanation of how the AI system works is not useful*

Types of Explanations

Fundamentally, an explanation is an answer to a question.[28] Using the question-driven framework for designing explanations, you can predict the type of questions users may ask of your AI system. The following are some of the most common types of questions:[29]

1. What did the system do?

2. Why did the system do it?

3. Why did the system not do this?

4. What would the system do if this happened?

5. How does it do it?

6. What is the overall model of how the system works?

7. What data does the system learn from?

8. How confident is the system about a prediction or an outcome?

9. What can I do to get a different prediction?

10. What changes are permitted to keep the same prediction?

You will need different types of explanations for different situations. This can depend on multiple factors like the type of user, what part of the user journey they are in, their level of expertise, industry type, or stakes of the situation. In this section, we will look at the following types of explanations and the questions they try to answer:

---

[28] Liao, Q. Vera, et al. "Questioning the AI: Informing Design Practices for Explainable AI User Experiences." ACM CHI Conference on Human Factors in Computing Systems (CHI 2020), 2021.

[29] Kaushik, Meena, et al. UXAI. uxai.design, 2020, www.uxai.design/.

1. Data use explanations

2. Descriptions

3. Confidence-based explanations

4. Explaining through experimentation

5. No explanation

Data Use Explanations

Data use explanations tell users what data is used and how the AI system interacts with this data, which can help users build the right level of trust in the AI's output. Data use explanations often answer the following questions:

1. What data does the system learn from?

2. Why did the system do it?

3. How does it do it?

Guidelines for Designing Data Use Explanations

Data use explanations describe the kind of information and the method used to derive a particular AI output. Here are some guidelines to help you design better data use explanations:

1. **Explain what data is used.**

   Ensure that users are made aware of any data that is collected, tracked, or monitored and that it's easy for them to find out how the data is collected, whether via sensors, user-entered data, or other sources.[30]

---

[30] Kershaw, Nat, and C. J. Gronlund. "Introduction to Guidelines for Human-AI Interaction." Human-AI Interaction Guidelines, 2019, https://docs.microsoft. com/en-us/ai/guidelines-human-ai-interaction/.

It can also help to tell users the source of data so that they don't overtrust or distrust the system. For example, a navigation service suggesting the time to leave for a meeting can tell users that it derived this insight using data from their calendar, GPS, and traffic information.

2. **Explain how the data is used.**

   Your AI system will use data to provide a recommendation or make a decision. Users need to understand how the AI system uses the data. For example, a marathon training app that asks users to provide their location data explains that the AI will use this information to generate routes and calculate distance. Explaining how data is used can help build user trust.

3. **Explain what's important.**

   Providing a list of data sources or a dump of data is not understandable. Your data use explanations need to highlight the relevant data during an interaction. Sometimes, you can surface which data sources had the greatest influence on the system output. Identifying influential data sources for complex models is still a growing area of active research but can sometimes be done. In cases where it can, the influential feature(s) can then be described for the user in a simple sentence or illustration.[31] For example, an app that identifies cat

---

[31] pair.withgoogle.com, `https://pair.withgoogle.com/`.

breeds can show similar instances where it correctly identified the breed. In this case, it need not show you all samples across all breeds.

4. **Highlight privacy implications.**

Your AI systems should respect user privacy. People will be wary of sharing data about themselves if they are not confident of how their data is used, stored, and protected. Sometimes users can be surprised by their own information when they see it in a new context. These moments often occur when someone sees their data used in a way that appears as if it weren't private or when they see data they didn't know the system had access to, both of which can erode trust.[32] For example, a service that uses a person's financial data to predict loan eligibility without explaining how it has access to this data is not trustworthy. Privacy is a key pillar of building trust with users. Communicate privacy and security settings on user data. Explicitly share which data is shared and which data isn't.[33] For example, a social music streaming service that shares what you are listening to with your friends needs to communicate explicitly that your friends can see your listening activity. Additionally, users should also be able to opt out of sharing this information.

---

[32] pair.withgoogle.com, `https://pair.withgoogle.com/`.
[33] pair.withgoogle.com, `https://pair.withgoogle.com/`.

Types of Data Use Explanations

You can help build and reinforce user trust by enabling users to see what data is used and how it interacts with your AI system. Data use explanations can be categorized into three types:

1.   Scope of data use

2.   Reach of data use

3.   Examples-based explanations

Scope of Data Use

Scope refers to the type and range of data used to provide a result. When using scope-based explanations, show an overview of the data collected about the user and which aspects of the data are being used for what purpose.[34] Sometimes users may be surprised to see their information shown in the product; this can erode trust. To avoid this, explain to users where their data is coming from and how it is used. Your data sources need to be part of the explanation. For example, a personal assistant may offer to book a cab for your appointment. In this case, the personal assistant needs to communicate that it knows about your appointment because your calendar is linked and you have configured your preferred mode of transport as an Uber. However, remember that there may be legal, fairness, and ethical considerations for collecting data and communicating about data sources used in AI.[35] Sometimes this may be legally required in certain regions.

---

[34] Kore, Akshay. "Designing AI products." woktossrobot.com, 2020, `https://woktossrobot.com/aiguide/`.

[35] pair.withgoogle.com, `https://pair.withgoogle.com/`.

🚘 Time to leave notification

Scope of data explanation ———————

Type and range of data used.

Figure 4-10. *Scope of data use.* *The notification has an explanation of how the AI knows it is time to leave for a flight, for example, based on flight bookings on your calendar and current traffic data*

Reach of Data Use

Reach-based explanations tell users if the system is personalized to them or a device or if it is using aggregated data across all users.[36] Your sources of data need to be part of reach-based explanations. For example, when Spotify creates personalized playlists for users, the description often explains how it was generated: "Your top songs 2020—the songs you loved most this year, all wrapped up." When Amazon's recommendation system on a product page shows aggregated suggestions, the explanation also explains the logic for the recommendation in a user-friendly manner: "Customers who read this book also read..."

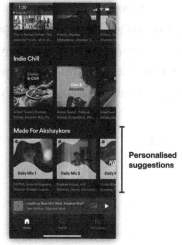

Figure 4-11. Reach of data use. (a) Amazon's shopping recommendations explain that the suggestions are aggregated across customers, that is, "Customers who bought this item also bought these." Source: www.amazon.in. (b) Spotify clarifies that the playlist suggestions are personalized, that is, made for the user. Source: Spotify app on iPhone

---

[36] pair.withgoogle.com, https://pair.withgoogle.com/.

Examples-Based Explanations

Sometimes it is tricky to explain the reasons behind an AI's output. In examples-based explanations, we show users examples of similar results or results from the training data relevant to the current interaction. Examples can help users build mental models about the AI's behavior intuitively. These explanations rely on human intelligence to analyze the examples and decide how much to trust the classification.[37] Examples-based explanations can be generic or specific.

Generic Explanations

The system shows users examples where it tends to perform well and where it performs poorly. For example, an AI that detects lung cancer from X-ray images can show the types of images it performs well on, which may have a proper orientation, may be well-lit, etc. It can also display images on which it performs poorly, which may not have a correct orientation, may have too many organs in view, etc.

---

[37] pair.withgoogle.com, https://pair.withgoogle.com/.

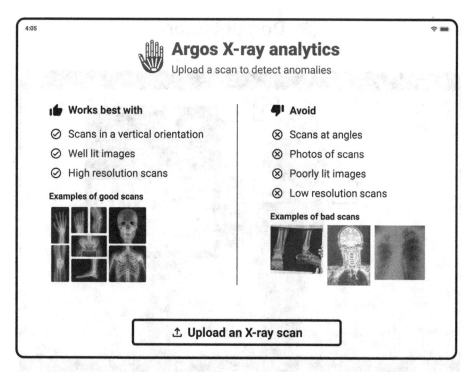

Figure 4-12. Generic examples-based explanation. X-ray analytics interface shows examples of images the system performs well on and where it doesn't

Specific Explanations

With specific examples-based explanations, the system displays the most similar examples relevant to the current interaction. For example, a dog classification AI might show similar images from its training data along with the prediction when you show it a photo of a poodle. Displaying similar images can help users judge whether they should trust the "poodle" classification.

🐾 Dog detector

Figure 4-13. Specific examples-based explanation. When it detects a dog breed, the dog classification system shows similar images from its training data along with the result. This helps users gauge how much they can trust the result. For example, if the similar images for a "poodle" prediction were photos of cats, you wouldn't trust the system's results

Descriptions

A description is a summary of how the AI system behaves or why it made a particular prediction. A description often answers the following question types:

1. What did the system do?

2. Why did the system do it?

3. What is the overall model of how the system works?

4. How does it do it?

Guidelines for Designing Better Descriptions

Letting users know the workings of your AI system in an understandable manner helps build trust and allows people to make appropriate judgments. Here are some guidelines that can help you design better descriptions:

1. **Explain the benefit, not the technology.**[38]

    As product creators, we are often excited about the underlying technologies, especially if we've solved a particularly hard problem. But most users don't care about the technology; they just want to get a job done. They don't necessarily need to understand the math behind an algorithm in order to trust it. Within the product experience, it is much better to help users understand how the technology benefits them. For example, in a spam filtering application, it is better to say "We help you remove spam from your inbox" than saying "We've created a sophisticated email filtering technique that categorizes email into spam and non-spam by using extensive training data."

    If you think your users would be interested in the underlying technology, you can always provide more details with tooltips and progressive disclosure. You can also provide this information in

---

[38] pair.withgoogle.com, https://pair.withgoogle.com/.

your marketing communications. If you talk about the AI system, focus on how it benefits the user and not how you built it.

2. **Make it understandable.**

Make sure that your descriptions are understandable to your users. Try to make your descriptions less like a user manual with technical jargon and more like an aid to decision-making. For example, typing "area of Poland" into the search engine Bing returns the literal answer (120,728 square miles) along with the note "About equal to the size of Nevada." The numeric answer is the more accurate, but the intuitive answer conveys the approximate size of Poland to far more people.[39]

3. **Account for situational tasks.**

It is essential to consider the risks of a user trusting an inaccurate suggestion. You can tailor the level of explanation by accounting for the situation and potential consequences. For example, in a logistics system, if the items being delivered are low stakes, like clothes or toys, you can get away with explaining the tentative time of delivery. However, suppose the items being delivered are high stakes, like surgical equipment or ventilators. In that case, you should let users know the tentative delivery time as well as limitations of your system, like the data refreshing

---

[39] Guszcza, Jim. "AI Needs Human-Centered Design." *WIRED*, 2018, www.wired.com/brandlab/2018/05/ai-needs-human-centered-design/.

every hour or the possibility of a delay. Additionally, you can give users more control by enabling them to contact the delivery agent or shipping provider.

4. **Explain what's important.**

   While interacting with your AI product, users may not need all the information. In fact, providing all the information might sometimes be detrimental. Explain only what's important in a given scenario, intentionally leaving out parts of the system's function that are highly complex or simply not useful.[40] For example, a user trying to select a movie to watch on an AI-based streaming service might find it irritating if the system overexplains its working. However, you can explain the system's entire workings through progressive disclosure or other channels like blog posts or marketing communications.

5. **Use counterfactuals** (sometimes).

   Using counterfactuals is a way of telling users why the AI did not make a certain prediction. They are the answers to questions like "What would happen if ___?" or "Why did the system not do this?" Counterfactuals are not always helpful, but sometimes they can significantly improve your explanations. For example, an AI service that predicts the selling price of your car can provide insights like "You would get 5% more if the paint was not chipped" or "You would get 10% less if you wait another three months." You need to determine if you should use counterfactuals through trial and error.

---

[40] pair.withgoogle.com, https://pair.withgoogle.com/.

Types of Descriptions

Designing user-centered explanations for your AI can significantly improve your product's user experience. In general, we can categorize descriptions into two types.

Partial Explanations

When deciding the description type during an interaction that can increase or maintain trust in your AI, a partial explanation is most likely the best one. In these descriptions, we intentionally leave out functions that are unknown, too complex to explain and understand, or simply not useful. In most cases, you can help build user trust without necessarily explaining exactly how an algorithm works or why it made a prediction. Your partial explanations can be generic or specific:

1. **Generic explanations**

   General system explanations talk about how the whole system behaves, regardless of the specific input. They can explain the types of data used, what the system is optimizing for, and how the system was trained.[41] For example, a marathon training app can say "This app uses your height, weight, and past runs to find a workout" when suggesting exercises.

2. **Specific explanations**

   They explain the rationale behind a specific AI output in a manner that is understandable. For example, a recipe recommendation system can say, "We recommended this recipe because you wanted to make a light meal with broccoli, pasta,

---

[41] pair.withgoogle.com, https://pair.withgoogle.com/.

and mushrooms," or a dog recognition system can say, "This dog is most likely a German shepherd because of XYZ features." Specific explanations are useful because they connect explanations directly to actions and can help resolve confusion in the context of user tasks.[42]

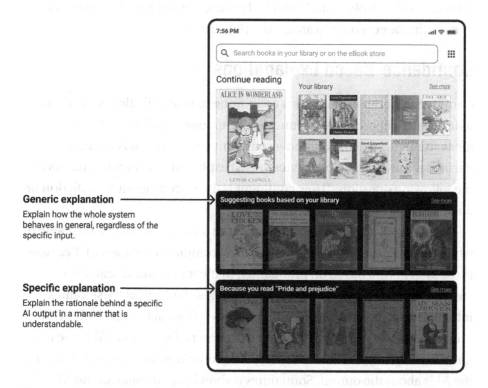

Generic explanation

Explain how the whole system behaves in general, regardless of the specific input.

Specific explanation

Explain the rationale behind a specific AI output in a manner that is understandable.

Figure 4-14. Examples of specific and generic explanations for book suggestions on an eBook reader

---

[42] pair.withgoogle.com, https://pair.withgoogle.com/.

Full Explanations

These are detailed explanations of how your AI system works or why it made a particular prediction. These can be in the form of research papers, blog posts, open source code, etc. In most cases, it is not advisable to give full explanations within your product's user experience. Such explanations might be irrelevant for the task and can also confuse users. If you really want to provide full explanations, you can do it through progressive disclosure, tooltips with external links, etc. The best place for full explanations can be on your product's blog, marketing collaterals, the company's landing page, or even the careers page to attract the right candidates.

Confidence-Based Explanations

Confidence-based explanations are unique to probabilistic systems like data science reports or AI products. A confidence level is a statistical measurement that indicates how confident the AI system is about a particular outcome.[43] Confidence-based explanations provide answers to the following question: "How confident is the system about a prediction or an outcome?"

Confidence is a readily available output from an AI system, the value of which ranges from 0 (no confidence) to 1 (complete confidence). Because these systems are based on probability, in most real-world scenarios, you will never get a definite 0 or 1. Sometimes, we convert these values into percentages. Displaying confidence levels can help users gauge how much to trust an AI's output. Rather than describing why the AI came to a particular decision, confidence-based explanations tell users how certain the AI is about the output. Sometimes it shows the alternatives the AI model considered.

---

[43] Kore, Akshay. "Designing AI products." woktossrobot.com, 2020, https://woktossrobot.com/aiguide/.

Let's assume that we build an AI system that detects donuts from images. We categorize results into high, medium, and low levels of confidence. A result with a value less than 40% has low confidence, while one with a value of more than 80% is categorized as high confidence, and anything between 40% and 80% has a medium level of confidence. The team building the AI decides these threshold values, which can vary across systems. Using these confidence levels and categories, we can customize the system's output. For example, if you show an image of a donut that results in a high level of confidence, we can design it to say, "That's a donut." If the confidence about the image is medium, then it can say, "Maybe that's a donut." When you show the donut detector a picture of pasta that results in a low confidence, it can say, "I don't think it's a donut."

Image	Confidence value	Confidence level	System output
	98%	High	"That's a Donut"
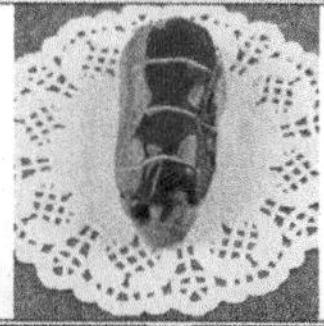	65%	Medium	"Maybe this is a Donut"
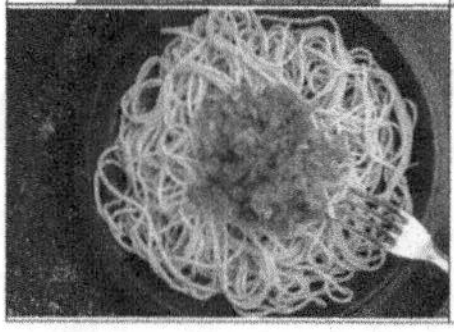	15%	Low	"Not a Donut"

Figure 4-15. *Examples of confidence levels and outputs for a system that identifies donuts*

Guidelines for Designing Confidence-Based Explanations

These explanations rely on the AI confidence values by indicating how certain the system is about the accuracy of results. Displaying confidence levels can help users build the right level of trust in the AI's outputs. The following are some guidelines that can help you build better confidence-based explanations:

1. **Determine if you should show confidence.**[44]

   Displaying confidence values can sometimes help users calibrate their trust in the AI's output. But confidence values are not always actionable. In many cases, it is not easy to make confidence values intuitive. Even if you're sure that your user has enough knowledge to properly interpret your confidence displays, consider how it will improve usability and comprehension of the system—if at all.[45] To avoid this, test with users if showing confidence is beneficial. To assess if showing confidence improves user trust and helps people make decisions, you can conduct user research on your target stakeholders. There is always a risk wherein showing confidence values can be irrelevant, distracting, or confusing and even have negative consequences.

   Here are some cases when you should **not** indicate confidence:

   a. Consider not showing the confidence value if it doesn't help with decision-making.

---

[44] pair.withgoogle.com, `https://pair.withgoogle.com/`.

[45] pair.withgoogle.com, `https://pair.withgoogle.com/`.

b. Don't show the confidence value if it misleads
    users, causes confusion, or creates distrust. An
    inaccurate high confidence result can cause users
    to trust an AI decision blindly. If the confidence
    value is misleading for certain users, reconsider how
    it is displayed, explain it further, or consider not
    showing it at all.

2. **Optimize for understanding.**

    Showing a confidence value should help users in
    making a decision and calibrate the right level of
    trust. Users can't always gauge if a 75% confidence
    is high or low or enough to make a decision.
    Using friendly terms like high or low can be more
    beneficial than saying 95% confidence. Sometimes,
    showing too much granularity can be confusing.
    Which option should the user choose between
    84.3% and 85% confidence? Is the 0.7% difference
    significant?

    Statistically, information like confidence scores can
    be challenging for different users to understand.
    Because different users may be more or less familiar
    with what confidence means, it is essential to
    test your confidence displays early in the product
    development process. We can display confidence
    levels in many ways. You can choose to reinterpret
    confidence into statements like "I'm not sure if
    this is a donut," "I'm pretty sure this is a donut," or
    "Maybe this is a donut" for a low, high, or medium
    confidence, respectively. Sometimes you can modify
    the user interface based on the level of confidence.

157

3. **Balance the influence of confidence levels on user decisions.**

   Product teams need to understand how much influence displaying confidence in the AI suggestion has on the user's decision. If showing confidence does not significantly impact the user's decision-making, presenting it might not make sense. On the other hand, if confidence has a disproportionate influence, it can lead to overtrust. Your confidence-based explanations should help users calibrate the right level of trust in the AI.

Types of Confidence-Based Explanations

Displaying confidence can help users gauge how much trust to put in an AI's output. If your research confirms that showing confidence improves decision-making, the next step is to choose the best type of explanation. Confidence-based explanations can be categorized into four types:

1. Categorical

2. N-best results

3. Numeric

4. Data visualizations

Categorical

Instead of showing a numerical value of the confidence (between 0–1), confidence levels are categorized into buckets like high/medium/low, satisfactory/unsatisfactory, etc. Generally, the product team determines the threshold confidence values and cutoff points in consultation with ML counterparts. It is important to think about the number of categories and what they mean. Determining categories will require trial and error and testing with your users. You can further use the category information to render the appropriate user interface, alter messaging, and indicate further user action. For example, when you ask a voice-enabled speaker to play a song, it could respond differently depending on its confidence in understanding your command. It might play the song for high-confidence cases, while it might ask you to repeat the command for low-confidence cases. Similarly, a search engine might tailor its UI based on its confidence about a query.

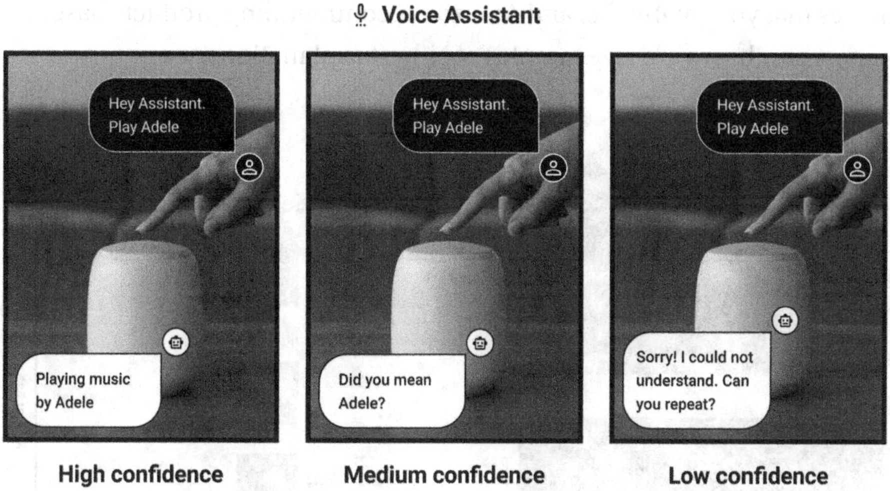

Figure 4-16. Categorical confidence. Based on the confidence level, the voice assistant provides different responses

N-Best Results

N-best means showing a specific or "n" number of best results that your AI outputs. We display multiple results in this method rather than presenting only one result with a particular confidence level. The confidence values of those results may or may not be indicated. This type of explanation is advantageous in low-confidence situations or in cases where it is valuable to provide users with alternatives. Showing multiple options prompts the user to rely on their own judgment. It also helps people build a mental model of how the system relates to different options.[46] Determining the number of alternatives will require testing and iteration.

Displaying N-best results is very common. For example, a smartphone keyboard with predictive text capabilities provides multiple choices of words. An application that detects animals from photos might show numerous options if it is unsure of what it's seeing, for example, this photo may contain a goat, llama, or ram. The autosuggest capability in Google Search that suggests possible queries, Netflix recommending multiple movies that you might like, and Amazon recommending products based on past purchases are all examples of N-best explanations.

Figure 4-17. *N-best results. (a) Predictive text on Gboard provides multiple response choices. (b) Google autosuggest suggests multiple search queries. Source: Google Search on Chrome. (c) Search result for the query "science fiction" on Audible provides multiple options to choose from. Source: Audible website on desktop*

---

[46] pair.withgoogle.com, https://pair.withgoogle.com/.

Numeric

In this method, numeric values of confidence are shown, often in the form of percentages. However, using numeric confidence explanations is risky since it assumes that users have a baseline knowledge of probability and an understanding of threshold values. Showing numeric explanations can even confuse users. It is difficult to gauge if an 87% confidence value is low or high for a particular context. Users might trust an AI that predicts a song they like with 87% confidence, but they would be wary of trusting it to drive their kids to school.

Make sure to provide enough context to users about your numeric explanations and what they mean. Remember that since AI systems are probabilistic, you will almost never get a 100% confidence value. There are two types of numeric explanations:

1. Specific explanations provide users with a prediction along with the AI's confidence value. For example, when you select a movie on Netflix, it shows a percentage match indicating your likelihood of enjoying it.

2. General explanations present an average confidence of the system, for example, this app recognizes dogs with a 90% accuracy.

(a) Match score on Netflix titles   (b) Confidence score on Photofeeler

Figure 4-18. Numeric confidence. (a) Percentage match score on Netflix. Source: Netflix website on desktop. (c) Photofeeler provides automated numeric scores (out of 10) on a profile picture. Source: www.photofeeler.com/

Data Visualizations

When it comes to indicating confidence values, data visualizations are graphical indications of certainty over a span of time, types of results, or any other metric. For example, a stock price or a sales forecast could include elements indicating a range of possible outcomes based on your AI's confidence. Your data visualizations can be static or interactive. Keep in mind that many data visualizations are best understood by expert users in specific domains. However, it is safe to assume that many people understand common visualizations like pie charts, bar graphs, trend lines, etc. Using a visualization and choosing the right type will require user research, testing, and iterations.

Using data visualizations to display confidence can be especially useful if you are designing AI systems for expert users. This type of user might appreciate your AI's understanding of their domain and trust it more.

***Figure 4-19. Data visualizations in Google Ads Keyword
Planner.*** *Data visualizations can sometimes be complex and are best
understood by expert users. Source:* `https://ads.google.com/`

Explaining Through Experimentation

People sometimes learn by tinkering with a system. In this method, you
can explain the AI and help users build a mental model of the system
by experimenting with it. By trying out the AI on the fly, people can
understand the system's behavior, strengths, and weaknesses and test its
limits. For example, a user might ask a smart speaker to play music that it
should be able to do or ask it to assemble their furniture, which might not
be possible. Experimentation can also be an opportunity to teach users
how to use your feature. This type of explanation often helps answer the
following questions:

1. What would the system do if this happened?

2. What can I do to get a different prediction?

3. What changes are permitted to keep the same
   prediction?

163

People are often curious about how an AI-powered feature will behave. Users can be impatient and might want to jump into your product experience right away. They might skip any onboarding flows you might've designed to explain the AI system. It can help keep your onboarding short and suggest low-risk, reversible actions they can try out. For example, a document scanning app might ask the user to take a picture of a receipt and convert it to a scan with selectable text. It might allow them to copy text from the image, edit the scan area, apply image filters, etc. In this case, trying out the image-to-text extraction feature is much better than explaining through words. Such small, contained experimentation environments can help users build a mental model of your system, and you can start building user trust.

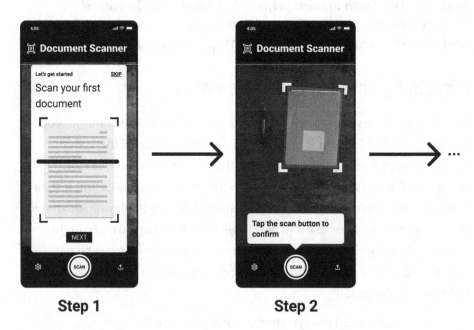

Step 1   **Step 2**

*Figure 4-20. **Experimentation**. Example of a document scanning app that allows users to try-out the feature during onboarding*

You need to be intentional about letting users interact with the AI and test its limits. Building an experimentation-based explanation can be time-consuming and require a lot of effort. Developing such experiences will need multiple rounds of user testing.

Guidelines to Design Better Experimentation Experiences

Allowing users to try out and tinker with your AI product can help build trust and improve usability. Users can get up to speed with your product faster. The following are some guidelines to help you design better experimentation-based explanations:

1. **Allow experimentation with specific features.**

   Asking users to try out the entire system as a way of experimentation is not explainable. Doing this can even confuse users about your product's value. You need to tie your explanations to a specific output. For example, changing the lighting in a document scanning application can give a better outcome, or in a search engine, modifying your search query can provide better results. Point users to specific features where they can quickly understand the value of your product. Otherwise, they may find the system's boundaries by experimenting with it in ways it isn't prepared to respond. This can lead to errors, failure states, and potentially erosion of trust in your product.[47]

---

[47] pair.withgoogle.com, https://pair.withgoogle.com/.

2. **Consider the type of user.**

   A user's willingness to experiment will depend on their goals. For example, an average consumer might enjoy trying out different AI-enabled features on a smart TV as part of the explanation, but a person buying hundreds of TVs for an enterprise might find this frustrating.

3. **Build cause-and-effect relationships.**

   People learn faster when they can build a cause-and-effect relationship between their actions and AI's outputs. Within an experimentation environment, a specific user action should generate a response. The AI can respond to the success or even failure of the system in the form of errors. Providing no response can keep users waiting and frustrated and can erode trust in your system. For example, a smart speaker that offers no response and keeps the user hanging despite an error is worse than providing an error response.

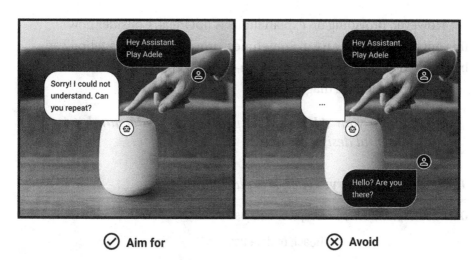

Figure 4-21. *Don't leave the user hanging.* (*Left*) *Provide an error response to let users know what happened and what they can do about it.* (*Right*) *Providing no response creates uncertainty and can erode trust*

No Explanation

In some cases, there is no benefit in explaining the AI's function. If the way an AI works fits a common mental model and matches user expectations for function and reliability, then there may not be anything to explain in the interaction.[48] For example, in a camera application that automatically adjusts to external lighting, it might be irrelevant to explain every time it adjusts the image. Your users might already have a mental model of when and how that happens. Explanations can sometimes get in the way of a user's actions. In the case of an AI writing assistant, it can be distracting if the system explains every time it corrects a word while you are drafting an email. It would also be wise to avoid explanations if they reveal private information or proprietary techniques.

---

[48] pair.withgoogle.com, https://pair.withgoogle.com/.

However, providing no explanations is not ideal. Consider how this might impact user trust. Sometimes, providing an explanation can be required by law, especially for high-stakes scenarios like criminal sentencing or medical diagnosis.

Explaining your AI system so people can understand it is a fundamental design challenge.

Table 4-1. *Summary: Types of explanations*

Data use explanations	Scope of data use	
	Reach of data use	
	Examples-based explanations	Generic explanation
		Specific explanation
Descriptions	Partial explanation	Generic explanation
		Specific explanation
	Full explanation	
Confidence-based explanations	Categorical	
	N-best results	
	Numeric	
	Data visualizations	
Explaining through experimentation		
No explanation		

Evaluating Explanations

When designing for AI explainability, you need to assess if your explanations increase user trust or make it easier for users to make decisions. When embarking on a new project, you can evaluate your explanations internally within your team and later with users.

Internal Assessment

You can evaluate your explanations with your product managers, designers, machine learning scientists, and engineers on your team. You can conduct your assessment on the following points:

1. Consider if your type of explanation is suitable for the user and the kind of product.

2. Observe how your team members interact with the explanation. Ask them what they understand from the explanation and what parts are confusing.

3. Determine if the components of your explanation are relevant to the user. Are we highlighting the right parts in the explanation?

4. Determine if your explanation has any implications on user privacy, proprietary information, or the product's security.

User Validation

You should also validate your explanations with the users of your product. Your user group should reflect the diversity of your audience. You can use qualitative or quantitative methods to validate your explanations.

Qualitative Methods

You can use different methods to validate your explanations qualitatively. While qualitative methods can be subjective, they provide great insight into how users perceive your explanations and if they are helpful.

1. **User interviews**

   You can conduct one-on-one interviews with users and ask them what they think about your explanation. Here, you need to check if their mental model matches your product's model of how the AI works.

2. **Surveys and customer feedback**

   You can float survey forms to your customers or ask for feedback inside the product while interacting with it. You might sometimes be able to validate your explanations by listening to customer service calls, reviews and feedback of your product on external websites and app stores, and public conversations about your product on social media.

3. **Task completion**

   You can ask your users to complete a predefined task on your product or a prototype and observe if your explanation helps them accomplish the task. You can also have a completion time defined as a metric of success. For example, an explanation is a success if the user is able to scan a document for the first time within one minute.

4. **Fly on the wall**

You can be a fly on the wall, that is, you can ask to be a silent observer on your user interactions. In many cases, your product might already be tracking user interactions on the product. Make sure to get the appropriate permissions when shadowing users. This method can help you uncover confusing parts of your product, where users hesitate, where you need to explain better, and which parts need further explanations.

Quantitative Methods

You can sometimes validate your explanations using quantitative methods like product logs, funnel diagrams, usage metrics, etc. However, quantitative methods will most likely provide weak signals and ambiguous information about explainability. For example, a search result page that indicates the confidence of results may have many drop-offs. But it is difficult to pinpoint with only quantitative data if the drop-off happened due to poor explainability or something else. Quantitative methods are suitable for finding broad problems in the product. They are a good starting point, but they need to be coupled with qualitative assessments.

Control

AI systems are probabilistic and will sometimes make mistakes. Your users should not blindly trust the AI's results. They should have the ability to exercise their own judgments and second-guess your AI's predictions. An AI that makes an incorrect prediction and does not allow for any other choice is not useful or trustworthy. For example, a music recommendation system that forces users to only listen to songs it suggests is not desirable. Giving users some control over your AI's algorithm and results will help them gauge the level of trust they should place in the AI's predictions. This makes them more likely feel the algorithm is superior and more likely continue to use the AI system in the future.[49] You can give users more control by allowing users to edit data, choose the types of results, ignore recommendations, and correct mistakes through feedback.

> *Users should have the ability to exercise their own judgments and second-guess your AI's predictions. They should not blindly trust the AI's results.*

Guidelines for Providing User Control

Providing control assures users that if things go wrong, they can always correct them. Incorporating control mechanisms like the ability to edit data, choose or dismiss results, and provide feedback allows users to assess the results and make their own judgments. Control mechanisms can also help your team improve the AI through feedback. Here are some guidelines that can help you design better control mechanisms.

Balance Control and Automation

Your AI system will not be perfect for all users all the time. You need to maintain a balance between AI automation and user control. When the AI makes mistakes or its predictions are slightly off, allow your users to adapt the output as needed. They may use some part of your recommendation, edit it, or completely ignore it. The amount of automation or control you provide can depend on the user's context, task, or industry. Think about your user's expectation of control over the system. Here are some things to keep in mind:

---

[49] Wilson, H. James, and Paul R. Daugherty. *Human + Machine: Reimagining Work in the Age of AI*. Harvard Business Review, 2018.

1. **Stakes of the situation**

   The level of control you provide will depend on
   the stakes of the situation. You would need to
   provide stakeholders with greater control in high-
   stakes scenarios like surgery, healthcare, finance,
   or criminal justice. A higher level of control can
   include greater visibility into the algorithms, the
   ability to edit the AI algorithm or underlying training
   data, or correcting AI's mistakes. On the other hand,
   users might be okay with lesser control in low-stakes
   situations like music recommendations, applying
   filters on photos, etc.

2. **Time required for AI to learn**

   Consider how long it would take for your AI to
   reach the target level of accuracy and usefulness. In
   many cases, your AI might learn user preferences
   from scratch. In the beginning, when it might
   not be as accurate or helpful, you can provide
   a greater level of user control by putting higher
   weightage on control mechanisms in the interface.
   Over time as the AI improves, you can shift the
   weightage to predictions. For example, in a movie
   recommendation service, in the beginning, a user
   may be asked to choose most of the titles from
   predefined categories like action, comedy, horror,
   etc., thereby providing more control over the results.
   As the system learns the user's preferences and
   reaches a level of acceptable accuracy, you can start
   showing AI recommendations—slowly increasing
   the weightage from predefined categories to
   personalized ones.

3. **Time required for users to learn**

   In cases where you're introducing new mental
   models, it might take time for users to learn the
   system's workings. Users might need time to
   understand how the AI system generated its results
   and the relationship between their input and AI's
   output. Your users may be wary of trusting the AI's
   recommendations from the start; they might tinker
   and experiment with the product before establishing
   the right level of trust. Even if your AI is highly
   accurate, it is a good idea to provide users with a
   greater level of control while they learn the system's
   mental model. For example, workers in a factory
   using a robot may choose to control it manually in
   the beginning and slowly hand off various tasks as
   they gain more trust.

Hand Off Gracefully

When your AI makes mistakes, the easiest way is to return control to the
user. It should be easy for users to take over the system. The level of control
and override would depend on the type of situation. Sometimes, it is risky
to hand off to a user immediately. For example, it can be dangerous if a self-
driving car suddenly asks the driver to take over at high speed. You need
to ensure that the handoffs are graceful. When this AI-to-manual control
transition happens, it's your responsibility to make it easy and intuitive
for users to pick up where the system leaves off quickly. That means the
user must have all the information they need in order to take the reins:
awareness of the situation, what they need to do next, and how to do it.[50]

---

[50] pair.withgoogle.com, `https://pair.withgoogle.com/`.

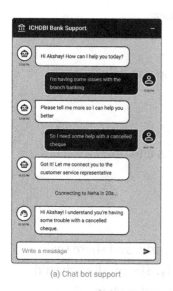

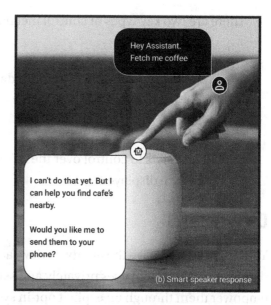

*Figure 4-22. **Hand off gracefully.** (a) Example of a banking chatbot that gracefully hands off to a human agent when it can't answer a specific question. (b) Voice assistant handing off control to the user when it can't do an action*

Types of Control Mechanisms

Control mechanisms allow users to take over the system when it makes mistakes and use their own judgment. Allowing users to ignore your AI's results, edit them, or give feedback makes it more likely for them to trust the system and continue using it. You can build different types of control mechanisms within your product. These may or may not exist together.

There are two key types of control mechanisms:

1. **Data control**

   This refers to control over input data that you or your users give the AI.

2. **Control over AI output**

   This refers to control over the AI's results and how they are displayed.

Data Control

Your users should have the ability to view, access, edit, and share their data that the AI system uses privately and securely. The product should empower them through an explicit opt-in system and explainable information of how the AI stores and uses their data. You can allow users to select the data that the system uses and disable parts of the data they don't want to be used. In some cases, this might result in limited functionality, inaccurate results, or even the AI not working. Communicate these limitations, but don't force users into providing data if they wish to opt out. Not allowing users control over their data erodes their trust in the system.

When collecting data, a best practice, and a legal requirement in many countries, is to give users as much control as possible over what data the system can use and how the AI can use this information. You may need to provide users the ability to opt out or delete their account. Ensure your system is built to accommodate this.[51]

> *Not allowing users control over their data erodes their trust in the system.*

---

[51] pair.withgoogle.com, https://pair.withgoogle.com/.

Data control is different from the digital consent that users provide when a product accesses their data. Terms and conditions or privacy policies are primarily designed to provide legally accurate information regarding the usage of people's data to safeguard institutional and corporate interests while often neglecting the needs of the people whose data they process. "Consent fatigue," the constant request for agreement to sets of long and unreadable data handling conditions, causes a majority of users to simply click and accept terms in order to access the services they wish to use. General obfuscation regarding privacy policies and scenarios like the Cambridge Analytica scandal in 2018 demonstrate that even when individuals provide consent, the understanding of the value regarding their data and its safety is out of an individual's control.[52]

Data control is different from digital consent.

Here are some considerations when designing data control mechanisms.

Global Controls

Provide global controls by allowing users to customize what the system monitors and how it behaves.[53] Here are some guidelines to design better global controls:

1. Let users enable or disable external inputs. For example, enabling or disabling location input in a shopping application can enable or disable a "trending near you" feature.

---

[52] Ethically Aligned Design: A Vision for Prioritizing Human Well-Being with Autonomous and Intelligent Systems. First ed., IEEE, 2019.

[53] Kershaw, Nat, and C. J. Gronlund. "Introduction to Guidelines for Human-AI Interaction." Human-AI Interaction Guidelines, 2019, https://docs.microsoft.com/en-us/ai/guidelines-human-ai-interaction/.

2. Allow users to indicate preferences. For example, a running app can allow users to indicate if they want to avoid slopes, rocky trails, or traffic.

3. Continue to clearly communicate permissions and settings throughout the usage of your AI. Ask permissions early in the user journey. Think about when the user might want to review preferences they've set in the past and consider reminding them of these settings when they shift into different contexts and may have different needs. They may also forget what they're sharing and why, so explain the reasons and benefits.[54]

4. Allow users to deny service or data by having the AI ask for permission before an interaction or providing the option during an interaction. Privacy settings and permissions should be clear, findable, and adjustable.[55]

5. Make sure that your controls are not too broad. Simply enabling or disabling an AI feature is a very broad control that leads to your users either using the product or not, that is, they need to fully trust your AI or not at all. For example, an AI-based music recommendation service with controls that only allow enabling or disabling the service would not be useful for many users. Users who want to use the service will be forced to accept all data inputs even if they aren't comfortable sharing parts of their data. Forced data sharing can erode your user's

---

[54] pair.withgoogle.com, `https://pair.withgoogle.com/`.
[55] "IBM Design for AI." ibm.com, `www.ibm.com/design/ai/`.

trust in the AI and easily make them susceptible to switching to a competitor. However, if users can control which parts of their data are shared with the service, they can calibrate their level of control and trust in the system. For example, enabling users to disable location data in the music recommendation service can disable location-based playlists but allow them to use the rest of the application. Keep in mind that there is a risk of being too granular with the level of control you provide. Users can be confused if they have to choose granular controls like the precision and recall of your ML models.

6. Don't stop users from deleting or modifying information that the system has already collected. For example, a search engine should allow users to delete previous searches, or a running app should allow users to remove past runs. Users should always maintain control over what data is being used and in what context. They can deny access to personal data that they may find compromising or unfit for an AI to know or use.[56] In many cases, this might be a legal requirement.

---

[56] "IBM Design for AI." ibm.com, www.ibm.com/design/ai/.

(a) iOS settings

(b) Nike Run App

Figure 4-23.   Global controls. (a) Data permissions and control in iOS settings. Source: Apple iOS 15. (b) Nike Run Club app settings allow users to modify data preferences. Source: Nike Run Club on iPhone

Editability

Your user preferences may change over time. Consider giving users the ability to adjust their preferences and data use at any point in time. Even if your AI's recommendations were not relevant initially, they might become better later as it gathers more information about the users' preferences. Your product should allow for people to erase or update their previous selections or reset your ML model to the default, non-personalized version.[57] For example, a movie recommendation service might adapt to a user's preferences over time, but it might not be as expected. As the user's tastes in movies evolve, they might want to change the preferences that your AI captured previously. Editability allows them to calibrate their recommendations and thereby their trust in the system.

(a) iOS settings for Nike Run App   (b) Profile settings in Nike Run App

Figure 4-24.   Editability of data. (a) iOS settings allow users to adjust preferences and how a specific app should use location data. Source: Apple iOS 15. (b) Nike Run Club app settings allow users to modify user profile information. Source: Nike Run Club on iPhone

---

[57] pair.withgoogle.com, https://pair.withgoogle.com/.

Removal and Reset

Tell users if they can remove their entire data or a part of it. Removal of data is different from disabling inputs. In the case of removal, all or a selection of previous data that the system collected is deleted. Additionally, you can allow users to reset their preferences entirely. The ability to reset is useful when your system provides a large number of irrelevant suggestions and the best way is to start from scratch. For example, a user logging into an online dating service might find the recommendations personalized for their past self, irrelevant after a few years. Their preferences for dates or partners might have changed. The current suggestions may not be desirable and can erode the user's trust in the service. While this system can painstakingly calibrate its recommendations over time, in such cases, resetting data is the best course of action.

(a) Removal of past searches in Google search

(b) Resetting YouTube activity data

Figure 4-25. Removal and reset. *(a) Google Search allows users to remove recently searched queries easily. Source:* `www.google.com/`. *(b) Resetting YouTube history data on the Google account dashboard. Source:* `https://myaccount.google.com/data-and-privacy`

Opting Out

Your system should allow users to opt out of sharing certain aspects of their data and sometimes all of their data. Opting out of sharing their entire data can be equivalent to a reset. Consider allowing users to turn off a feature and respect the user's decision not to use it. However, keep in mind that they might decide to use it later and make sure switching it on is easy. While users may not be able to use the AI to perform a task, consider providing a manual, nonautomated way to complete it.

Empower users to adapt your AI output to their needs.

(a) Contact permissions on iOS

(b) Whatsapp permissions in iOS settings

Figure 4-26. Opting out. *(a) iOS settings allow users to opt out of providing contact data to certain apps. Source: Apple iOS 15. (b) Users can opt out of giving permissions in iOS settings for WhatsApp. Source: Apple iOS 15*

Control over AI Output

Your AI won't be perfect for every user all the time. Empower users to adapt your AI output to their needs, edit it, ignore it, or turn it off. Here are some considerations when enabling users to control the AI's outputs.

Provide a Choice of Results

If possible, allow users to select their preferred recommendations from a number of results, for example, a movie recommendation service that shows suggestions but allows users to choose which title to watch. You can ask users to pick which results they would like to see more, which can help the AI model improve over time. For example, a news app allows users to choose the topics they'd like to see more. Keep in mind that providing multiple recommendations is not always possible.

Figure 4-27. *Choice of results. (a) Netflix allows users to choose a title from their personalized suggestions. Source:* www.netflix.com/. *(b) Users can choose from articles recommended for them on Medium. Source:* https://medium.com/. *(c) Amazon shows multiple choices of results in its book suggestions. Source:* https://amazon.in/

Allow Users to Correct Mistakes

Any AI system will inevitably be wrong. Your product can empower users to correct the AI's mistakes through feedback. You might allow users to configure a list of results and their ranking, up-vote or down-vote results, or even provide detailed feedback. For example, a service that creates automatic playlists can allow users to like or dislike a song, change the order of songs, add or delete titles, etc. You can sometimes enable users to undo the inferences made by the system. Allowing a user to correct mistakes can establish the mental model that your system improves over time, and the user can play a part in this improvement. Being able to correct mistakes helps users to calibrate their trust over time.

(a) Netflix thumbs down

(b) Apple Music feedback

(c) Youtube recommended videos

Figure 4-28. *Correcting AI mistakes.* *(a) Users can provide feedback on Netflix titles recommended to them. Source:* www.netflix.com/. *(b) Apple Music allows users to correct recommendations on its personalized playlists by liking and disliking songs or changing the order of tracks. Source: Apple Music app on iPhone. (c) YouTube allows users to correct its recommendations by providing feedback mechanisms. Source:* https://youtube.com/

Support Efficient Dismissal

Make it easy to dismiss any irrelevant or undesired AI recommendations. Make sure to clarify how to dismiss a result, for example, swipe left on a dating app, saying "Cancel" in the case of a voice assistant, hiding or reporting ads in a search engine, etc. Don't obfuscate the method of dismissal, for example, a dismiss button for a text autocomplete feature that is too small or not obvious enough.[58]

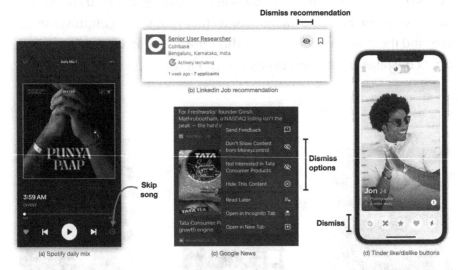

Figure 4-29. Support efficient dismissal. *(a) Users can easily skip songs on personalized daily playlists on Spotify. Source: Spotify app. (b) Google News allows users to dismiss results on its news feed. Source: Google News on Chrome. (c) Tinder app has an easy mechanism of swiping left for dismissing profile suggestions. Source:* `https://kubadownload.com/news/tinder-plus-free/`

---

[58] Kershaw, Nat, and C. J. Gronlund. "Introduction to Guidelines for Human-AI Interaction." Human-AI Interaction Guidelines, 2019, `https://docs.microsoft.com/en-us/ai/guidelines-human-ai-interaction/`.

Make It Easy to Ignore

Make it easy to ignore irrelevant or undesired AI recommendations. The ability to ignore a recommendation is especially useful in cases where the AI has a low level of confidence. You can do this by providing your recommendations unobtrusively. For example, a shopping application can show recommended products below the fold of the main product page that are easy to ignore.

People trust things that other people have trusted.

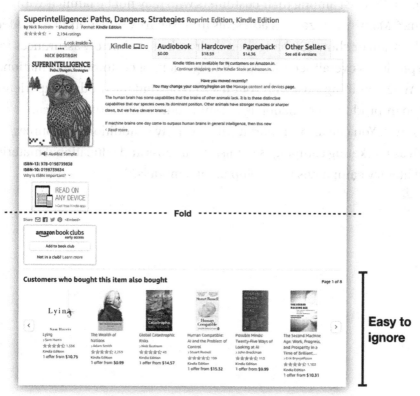

Amazon product page with recommendations

Figure 4-30. *Make it easy to ignore. Amazon recommendations on its website appear below the fold making them easy to ignore. Source:* `https://amazon.in/`

Borrowing Trust

People trust things that other people have trusted. You might trust buying a car from Tesla if you know your friends have bought it. Through social proof, you can borrow trust from your customers, peers, service providers, reputable brands, or trustworthy organizations. For example, The Currency Shop, an Australian currency comparison site, strategically displays other well-known and recognizable brands in Australia such as ANZ, OFX, and Commonwealth Bank within their landing page, which helps build trust among their customers who may not be familiar with them.[59] Many organizations add seals from security providers like McAfee and Norton or show SSL badges on their website to borrow assurances for the product's overall security and management of customer information.

When building trust with your user, go beyond the product experience. When in-product information is not sufficient to build trust, you can borrow it. You can also support it with a variety of additional resources, such as marketing campaigns to raise awareness and educational materials and literacy campaigns to develop mental models.[60]

---

[59] Nasir, Amirul. "8 UX Surefire Ways to Design for Trust." Medium, 2017, https://uxdesign.cc/8-ux-surefire-ways-to-design-for-trust-7d3720b57a4c.

[60] pair.withgoogle.com, https://pair.withgoogle.com/.

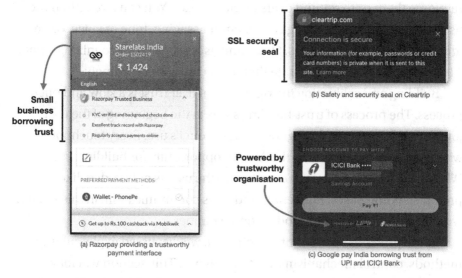

(a) Razorpay providing a trustworthy payment interface

(b) Safety and security seal on Cleartrip

(c) Google pay India borrowing trust from UPI and ICICI Bank

Figure 4-31. *Borrowing trust. (a) Starelabs website borrows trust from Razorpay in its payment interface. Source:* `www.starelabs.com/`. *(b) SSL security seal is a visual indicator that lets Cleartrip's visitors know that the organization values online security and privacy. Source:* `www.cleartrip.com/`. *(c) Google Pay borrows trust from a government system and a large bank to indicate that using the service to make a transaction is safe. Source: Google Pay India app*

> *The process of trust-building starts at the beginning of the user journey and continues throughout your product's use.*

Opportunities for Building Trust

To build trust, you need to explain how your AI works and allow your users to be in control of their relationship with the system. Some products are used every day, so their mental model gets formed by ongoing use. But some products are only meant to be used occasionally. For these products, mental models might erode over time, so it's helpful to consider ways to reinforce them or to remind users of the basics.[61] You can strengthen the user's mental model by using consistent messaging, tone of voice, or an identity for AI-based features. Over time users can start recognizing what is AI-powered or not and calibrate their expectations accordingly.

Building and calibrating the right level of user trust is a slow and ongoing process. The process of trust-building starts at the beginning of the user journey and continues throughout your product's use. Every touchpoint where the user interacts with the AI is an opportunity for building trust. While trust-building can start even before the user is onboarded, like product landing pages, marketing, and sales communication, in this section, we will focus mainly on in-product opportunities.

In the previous sections of this chapter, we discussed explainability methods, control mechanisms, and their types. This section will look at how we can apply those in your AI interfaces to build and calibrate user trust.

The following are some key trust-building opportunities for your AI:

1. Onboarding

2. User interactions

3. Loading states and updates

4. Settings and preferences

5. Errors

Onboarding

Onboarding is the process of introducing a new user to a product or service. Your onboarding experience begins even before a user purchases or downloads the product. Users can start forming mental models of your product from your website, marketing communications, or even word of

---

[61] pair.withgoogle.com, https://pair.withgoogle.com/.

mouth. Onboarding should not only introduce the system but also set the users' expectations of how it will work, what it can do, and how accurate it is.[62] Here are some guidelines that can help you design better onboarding experiences.

Set the Right Expectations

Many AI products set users up for disappointment by promising "AI magic" that can lead to users overestimating the AI's capabilities. Though product developers may intend to shield users from a product's complexity, hiding how it works can set users up for confusion and broken trust. It's a tricky balance to strike between explaining specific product capabilities—which can become overly technical, intimidating, and boring—and providing a high-level mental model of your AI-powered product.[63] Here are some recommendations that can help set the right expectations of your AI:

1. Be upfront and make clear what your system can and can't do. You can also collaborate with your marketing team to define this messaging in-product and within marketing collaterals.

2. Make clear how well the system does its job. Help users understand when the AI might make mistakes. For example, a service that recognizes cats from images might not perform too well on jungle cats.

3. Set expectations for AI adaptation. Indicate whether the product learns over time. Clarify that mistakes will happen and that user input will teach the

---

[62] Fisher, Kristie, and Shannon May. "Predictably Smart—Library." Google Design, 2018, https://design.google/library/predictably-smart/.
[63] pair.withgoogle.com, https://pair.withgoogle.com/.

product to perform better.[64] Let users know that the
AI may need their feedback to improve over time.
Communicate the value of providing feedback.

4. Show examples of how it works to clarify the
value of your product. Explain the benefit, not the
technology.

Introduce Features Only When Needed

Onboard in stages and at appropriate times. Avoid introducing new
features when users are busy doing something unrelated. This is especially
important if you are updating an existing product or a feature that changes
the AI's behavior. People learn better when short, explicit information
appears right when they need it.[65] Introduce AI-driven features when it is
relevant to the user. Avoid introducing AI-driven features as a part of a long
list of capabilities.

---

[64] Kershaw, Nat, and C. J. Gronlund. "Introduction to Guidelines for Human-AI
Interaction." Human-AI Interaction Guidelines, 2019, `https://docs.microsoft.com/en-us/ai/guidelines-human-ai-interaction/`.

[65] pair.withgoogle.com, `https://pair.withgoogle.com/`.

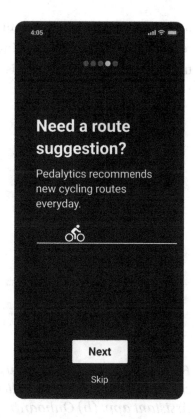

Figure 4-32.   Introduce features only when needed. (left) Aim to introduce AI-driven features only when it is relevant to the user. (right) Avoid introducing AI features as a part of a long list of capabilities

Clarify Data Use

As you onboard users to a new feature, they might have various concerns around privacy, security, and how their data is used. Make sure that users are aware of any data that is collected, tracked, or monitored and that it's easy for them to find out how the data is collected, whether via sensors, user-entered data, or other sources.[66] Ask for data permissions when relevant. For example, a running app might ask for location and GPS tracking permissions when it is creating route suggestions.

Figure 4-33. Clarify data use. *(a) Onboarding for Google Assistant summarizes how data is collected, shared, and used. Source: Google Assistant app. (b) Onboarding for the voice match feature on Google Assistant describes what data is used and how the model works. Source: Google Assistant app. (c) Apple Maps privacy policy has a clear and understandable explanation of data use. Source: Apple Maps app*

---

[66] Kershaw, Nat, and C. J. Gronlund. "Introduction to Guidelines for Human-AI Interaction." Human-AI Interaction Guidelines, 2019, `https://docs.microsoft.com/en-us/ai/guidelines-human-ai-interaction/`.

Allow Users to Control Preferences

Give users control over their data and preferences as they get started. Give them the ability to specify their preferences and make corrections when the system doesn't behave as expected, and give them opportunities to provide feedback.[67] This can also help set expectations that the AI will adapt to preferences over time.

(a) Preference selection on Headway app onboarding

(b) Spotify onboarding

(c) Preference selection on Flipboard

Figure 4-34.   Allow users to control preferences. *(a) Onboarding for the Headway app asks users to indicate their preferences. Source: Headway app. (b) Spotify onboarding with language preferences. Source:* https://medium.com/@smarthvasdev/deep-dive-into-spotifys-user-onboarding-experience-f2eefb8619d6. *(c) Preference selection in Flipboard's onboarding. Source: Flipboard app*

---

[67] pair.withgoogle.com, https://pair.withgoogle.com/.

Design for Experimentation

People sometimes skip the onboarding process because they are eager to use the system. Make it easy for users to try the experience first. Your onboarding can include a "sandbox" experience that enables them to explore and test the product with low risk or initial commitment.

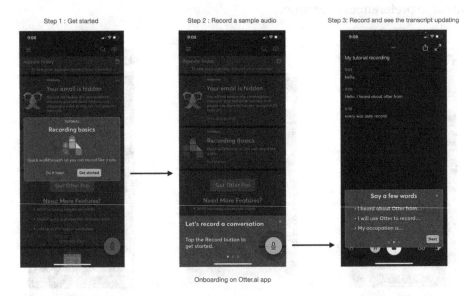

Onboarding on Otter.ai app

Figure 4-35.   Design for experimentation. Walk-through on the Otter.ai app to try out using the AI in a sandbox environment

Reboarding

Consider reboarding, that is, onboarding again, if there are noteworthy improvements in how a feature works or the data it uses. Ensure that these changes are significant enough for users to notice. You can also reboard existing users who are interacting with your system after a long time. In such cases, it is good to ask users if they want to be onboarded again and provide the option of skipping. Reboarding can also be used when a user forgets how to use your service and needs a refresher in the form of tutorials, wikis, etc.

User Interactions

Once your users are onboarded and have started using the system, you need to ensure that their trust is maintained and calibrated regularly. The user's interactions with your AI should strengthen established mental models. You need to set the right expectations of the AI, explain how the system works, and provide user control over preferences. You can help strengthen the mental models of your AI by maintaining a consistent messaging of its value, confidence, improvements over time, and data use. Here are some guidelines that can help you maintain and calibrate trust across the user journey.

Set the Right Expectations

You need to set the right expectations of your system's capabilities and performance. Interactions with your AI system should help your users in calibrating the right level of trust. Here are some recommendations:

1. Explain how your system works through partial descriptions.

2. Make sure that you explain what your AI feature can and can't do. Ensure that your users are aware of the full breadth of functionality. For example, a search functionality within a pharmacy website outlines the types of searches you can make.

3. Make it clear how well your system performs. Make sure your users understand when your AI makes mistakes. For example, a translation service built for American English speakers might not perform perfectly with a British accent.

4. You can use the AI's confidence levels to indicate performance and accuracy. This can enable users to calibrate the right level of trust.

5. Set expectations for adaptation. Let users know that the system improves over time through user feedback. Encourage users to provide feedback when appropriate.

Figure 4-36. Set the right expectations. (a) Search input on LinkedIn indicates the scope of the search on different pages. Source: www.linkedin.com/. (b) Netflix encourages users to provide feedback on recommendations. Source: www.netflix.com/

Clarify Data Use

Make sure your users understand what data is collected, tracked, and monitored by the system and how it is used. You should clarify which parts of their data are used for what purpose, what is private, and what is shared. You can also let your users know if the AI's results are personalized for them or aggregated over lots of users.

Build Cause-and-Effect Relationships

You can strengthen your AI's mental models by building cause-and-effect relationships between your user and the AI system. In many cases, the perfect time to show an explanation is in response to a user's action. People understand better if they can establish a relationship between their response and the AI's output. A user's relationship with your product can evolve over time through back-and-forth interactions that reveal the AI's strengths, weaknesses, and behaviors.

Allow Users to Choose, Dismiss, and Ignore AI Results

Users can build a more trustworthy relationship with your AI system if they can use their own judgment. Whenever possible, your AI system should provide a choice of results. If your AI's results are not helpful, users should be able to ignore or dismiss them easily.

Loading States and Updates

While system updates can improve the AI's performance, they may also lead to changes that are at odds with the user's current mental model.[68] Inform your users about any changes to your AI's capabilities or performance. Loading states and updates are good opportunities for informing your users when the AI system adds to or updates its capabilities. While the users are waiting, you can use the time to explain the new capability, recalibrate trust, or establish a new mental model. The following are some recommendations to help you design better loading states and update interfaces for your AI product:

---

[68] Fisher, Kristie, and Shannon May. "Predictably Smart—Library." Google Design, 2018, https://design.google/library/predictably-smart/.

1. Provide users with information about updates to the system, for example, a "What's new?" section to inform users about the AI system additions or capability updates.[69]

2. Inform users about any changes to the privacy policy or legal regulations.

3. Inform users about any changes to how data is used and shared.

4. Don't obfuscate updates in the algorithm, especially when they lead to direct changes in the AI's behavior, for example, ranking of results in a stock prediction service.

---

[69] Kershaw, Nat, and C. J. Gronlund. "Introduction to Guidelines for Human-AI Interaction." Human-AI Interaction Guidelines, 2019, https://docs.microsoft.com/en-us/ai/guidelines-human-ai-interaction/.

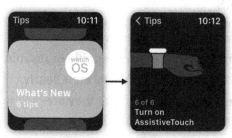

(a) What's New on Apple WatchOS

(b) Feature introduction on Apple Notes

(c) Microsoft Lens update

Figure 4-37. Inform users of changes to AI products. (a) "What's New" section on the Apple Watch. Source: watchOS 8. (b) New feature callout on Apple Notes app. Source: Apple Notes app. (c) New feature introduction on Microsoft Lens. Source: Microsoft Lens app

Settings and Preferences

Your AI product's settings and preferences section is an excellent destination for providing transparency and explaining how the system uses user data. It is also a great place to provide users control over their data and preferences. You can give users the ability to view, access, edit, and share their data that the AI system uses in a private and secure manner. Here are some guidelines that can help you design better settings interfaces for your AI products.

Provide Global Data Controls

Provide global controls by allowing users to customize any data that is collected, tracked, or monitored by the system. Let users enable or disable any external inputs like location, call information, contacts, etc. Clearly communicate permissions and privacy settings. Make sure they are findable and adjustable.

Clarify Data Use

In your system settings, make sure to communicate what data is collected, tracked, and monitored by the system and how it is used. You should clarify which parts of their data are used for what purpose, what is private, and what is shared.

Allow Editing Preferences

Your users might have provided their preferences during onboarding. Sometimes, your AI can learn user preferences over time. However, user choices can change over time. Allow users to specify or edit their preferences at any point in time. For example, a music recommendation service might adapt to a user's likings over time, but it might not be as expected. As users' tastes in music change, they might want to change the preferences that your AI captured previously.

Allow Users to Remove or Reset Data

Whenever possible, allow users to remove their entire data or a part of it. You can also provide users with the ability to reset their information and start from scratch. This is especially useful if the system has learned incorrect preferences leading to irrelevant predictions. In such cases, it is best to reset and start from scratch.

Allow Opting Out

You should allow users to opt out of the system even if it means that they can no longer use your service. Respect the user's decision not to use your service and communicate appropriately. Remember that building trust is a long process, and your users might decide to use the service at a later point. You can also enable users to opt out of sharing certain aspects of their data partially.

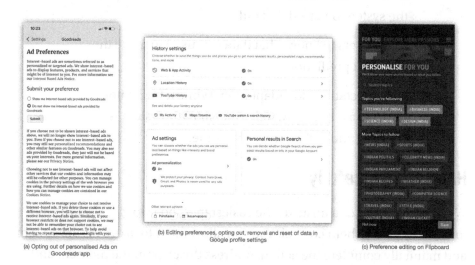

(a) Opting out of personalised Ads on Goodreads app

(b) Editing preferences, opting out, removal and reset of data in Google profile settings

(c) Preference editing on Flipboard

Figure 4-38. Settings and preferences. (a) Goodreads app allows users to opt out of personalized ads. Source: Goodreads app. (b) Editing preferences, opting out, removing data, and resetting data on the Google account dashboard. Source: https://myaccount.google. com/data-and-privacy. (c) Preference selection in Flipboard's settings. Source: Flipboard app

Errors

When your AI makes mistakes, your users might trust it less. However, errors are great opportunities to explain how the system works and why the AI made a mistake, collect feedback, and recalibrate user trust. Here are some suggestions for using errors as trust-building opportunities.

Adjust User Expectations

Because AI systems are probabilistic, they are inevitably going to make mistakes. When users run into an error, you can use this as an opportunity to adjust or set new user expectations. The following are some guidelines for adjusting user expectations:

1. Repair broken trust by making clear why the system made a mistake. Enable users to understand why the system behaved as it did.

2. Allow users to know what data was used to make the incorrect prediction.

3. Set expectations for adaptation. While it might not be accurate right now, tell users that the system is learning and will improve over time.

Hand Off Gracefully

When your AI makes a mistake, the easiest method is to give users a path forward by giving them control. You can allow them to take over the system and manually complete the action. Address the error in the moment and let users know how you will resolve the problem. However, you need to design the AI-to-manual transition carefully to ease your users into taking control of the system.

Allow Users to Correct AI Mistakes

You can enable users to correct the AI's mistakes through feedback. Providing feedback can also help set expectations for adaptation. Prevent the error from recurring: give users the opportunity to teach the system the prediction that they were expecting, or in the case of high-risk outcomes, completely shift away from automation to manual control.[70] Ability to correct mistakes helps users calibrate their trust in your AI system.

Allow Users to Choose, Dismiss, and Ignore AI Results

When the AI makes mistakes, users should be able to make their own judgments. If possible, allow users to choose from a number of results. When a result is irrelevant, users should be able to ignore or dismiss it. Allowing users to choose, dismiss, or ignore its results can help your AI build a more trustworthy relationship with your users.

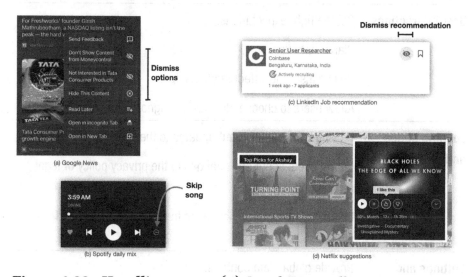

Figure 4-39. Handling errors. (a) Google News allows users to dismiss recommendations. Source: Google News on Chrome. (b) Users can easily skip songs on personalized daily playlists on Spotify. Source: Spotify app. (c) Hiding and dismissing irrelevant job opportunities on LinkedIn. Source: www.linkedin.com/. (d) Netflix encourages users to correct the AI's mistakes through feedback on recommendations. Source: www.netflix.com/

---

[70] pair.withgoogle.com, https://pair.withgoogle.com/.

Table 4-2. *Opportunities for building trust in your user journey*

Opportunities	Considerations
Onboarding	Set the right expectations.
	Introduce features only when needed.
	Clarify data use.
	Allow users to control preferences.
	Design for experimentation.
	Consider reboarding existing users periodically.
User interactions	Set the right expectations.
	Clarify data use.
	Build cause-and-effect relationships.
	Allow users to choose, dismiss, and ignore AI results.
Loading states and updates	Provide information about updates to the system.
	Inform users about any changes to the privacy policy or legal regulations.
	Inform users about any changes to how data is used and shared.
Settings and preferences	Provide global data controls.
	Clarify data use.
	Allow editing preferences.
	Allow users to remove or reset data.
	Allow opting out.

(*continued*)

Table 4-2. (*continued*)

Opportunities	Considerations
Errors	Adjust user expectations.
	Hand off gracefully.
	Allow users to correct AI mistakes.
	Allow users to choose, dismiss, and ignore AI results.

Personality and Emotion

Our emotions give cues to our mental states.

We tend to anthropomorphize AI systems, that is, we impute them with human-like qualities. Consumer demand for personality in AI dates back many decades in Hollywood and the video game industry.[71] Many popular depictions of AI like Samantha in the movie *Her* or Ava in *Ex Machina* show a personality and sometimes even display emotions. Many AI systems like Alexa or Siri are designed with a personality in mind. However, choosing to give your AI system a personality has its advantages and disadvantages. While an AI that appears human-like might feel more trustworthy, your users might overtrust the system or expose sensitive information because they think they are talking to a human. An AI with a personality can also set unrealistic expectations about its capabilities. If a user forms an emotional bond with an AI system, turning it off can be difficult even when it is no longer useful. It is generally not a good idea to

---

[71] Lim, Jeanne. "Personality In AI: Why Someday Soon Your Toaster Could Be an Extrovert." Hanson Robotics, 2019, www.hansonrobotics.com/personality-in-ai-why-someday-soon-your-toaster-could-be-an-extrovert/.

imbue your AI with human-like characteristics, especially if it is meant to act as a tool like translating languages, recognizing objects from images, or calculating distances.

> *While an AI that appears human-like might feel more trust-worthy, your users might overtrust the system.*

We're forming these tight relationships with our cars, our phones, and our smart-enabled devices.[72] Many of these bonds are not intentional. Some argue that we're building a lot of smartness into our technologies but not a lot of emotional intelligence.[73] Affect is a core aspect of intelligence. Our emotions give cues to our mental states. Emotions are one mechanism that humans evolved to accomplish what needs to be done in the time available with the information at hand—to satisfice. Emotions are not an impediment to rationality; arguably, they are integral to rationality in humans.[74] We are designing AI systems that simulate emotions in their interactions. According to Rana El Kaliouby, the founder of Affectiva, this kind of interface between humans and machines is going to become ubiquitous, that it will just be ingrained in the future human-machine interfaces, whether it's our car, our phone, or our smart device at our home or in the office. We will just be coexisting and collaborating with these new devices and new kinds of interfaces.[75] The goal of disclosing the agent's "personality" is to allow a person without any knowledge of AI technology to have a meaningful understanding of the likely behavior of the agent.[76]

---

[72] Ford, Martin R. *Architects of Intelligence*. Packt, 2018.

[73] Ford, Martin R. *Architects of Intelligence*. Packt, 2018.

[74] Ethically Aligned Design: A Vision for Prioritizing Human Well-Being with Autonomous and Intelligent Systems. First ed., IEEE, 2019.

[75] Ford, Martin R. *Architects of Intelligence*. Packt, 2018.

[76] Chia, Hui. "The 'Personality' in Artificial Intelligence." Pursuit, 2019, https://pursuit.unimelb.edu.au/articles/the-personality-in-artificial-intelligence.

Here are some scenarios where it makes sense to personify AI systems:

1. Avatars in games, chatbots, and voice assistants.

2. Collaborative settings where humans and machines partner up, collaborate, and help each other. For example, cobots in factories might use emotional cues to motivate or signal errors. An AI assistant that collaborates and works alongside people may need to display empathy.

3. If your AI is involved with caregiving activities like therapy, nursing, etc., it might make sense to display emotional cues.

4. If AI is pervasive in your product or a suite of products and you want to communicate it under an umbrella term. Having a consistent brand, tone of voice, and personality would be important. For example, almost all Google Assistant capabilities have a consistent voice across different touchpoints like Google Lens, smart speakers, Google Assistant within Maps, etc.

5. If building a tight relationship between your AI and the user is a core feature of your product.

Designing a personality for AI is complicated and needs to be done carefully.

Guidelines for Designing an AI Personality

Designing your AI's personality is an opportunity for building trust. Sometimes it makes sense to imbue your AI features with a personality and simulate emotions. The job of designing a persona for your AI is complicated and needs to be done carefully. Here are some guidelines to help you design better AI personas.

Don't Pretend to Be Human

People tend to trust human-like responses with AI interfaces involving voice and conversations. However, if the algorithmic nature and limits of these products are not explicitly communicated, they can set expectations that are unrealistic and eventually lead to user disappointment or even unintended deception.[77] For example, I have a cat, and I sometimes talk to her. I never think she is an actual human but is capable of giving me a response. When users confuse an AI with a human being, they can sometimes disclose more information than they would otherwise or rely on the system more than they should.[78] While it can be tempting to simulate humans and try to pass the Turing test, when building a product that real people will use, you should avoid emulating humans completely. We don't want to dupe our users and break their trust. For example, Microsoft's Cortana doesn't think it's human, and it knows it isn't a girl, and it has a team of writers that's writing for what it's engineered to do.[79] Your users should always be aware that they are interacting with an AI. Good design does not sacrifice transparency in creating a seamless experience. Imperceptible AI is not ethical AI.[80]

> *Good design does not sacrifice transparency in creating a seamless experience. Imperceptible AI is not ethical AI.*[81]

---

[77] pair.withgoogle.com, https://pair.withgoogle.com/.

[78] pair.withgoogle.com, https://pair.withgoogle.com/.

[79] Agrawal, Rajat. "Building Personalities for AI: In Conversation with Jonathan Foster and Deborah Harrison." Microsoft Stories India, 2019, https://news.microsoft.com/en-in/features/building-personalities-ai-jonathan-foster-deborah-harrison/.

[80] "IBM Design for AI." ibm.com, www.ibm.com/design/ai/.

[81] "IBM Design for AI." ibm.com, https://www.ibm.com/design/ai/.

Clearly Communicate Boundaries

You should clearly communicate your AI's limits and capabilities. When interacting with an AI with a personality and emotions, people can struggle to build accurate mental models of what's possible and what's not. While the idea of a general AI that can answer any questions can be easy to grasp and more inviting, it can set the wrong expectations and lead to mistrust. For example, an "Ask me anything" callout in a healthcare chatbot is misleading since you can't actually ask it anything—it can't get you groceries or call your mom. A better callout would be "Ask me about medicines, diseases, or doctors." When users can't accurately map the system's abilities, they may overtrust the system at the wrong times or miss out on the greatest value-add of all: better ways to do a task they take for granted.[82]

---

[82] pair.withgoogle.com, https://pair.withgoogle.com/.

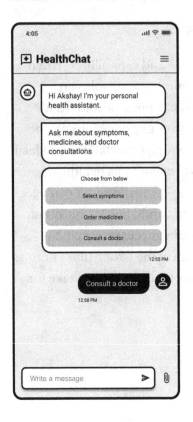

 **Aim for**

 **Avoid**

Figure 4-40.   Healthcare chatbot: Clearly communicate boundaries. (left) Aim to explain what the AI can do. In this example, the bot indicates its capabilities and boundaries. (right) Avoid open-ended statements. In this example, saying "Ask me anything" is misleading since users can't ask anything they want

Consider Your User

When crafting your AI's personality, consider whom you are building it for and why they would use your product. Knowing this can help you make decisions about your AI's brand, tone of voice, and appropriateness within the target user's context. Here are some recommendations:

1. Define your target audience and their preferences. Your user persona should consider their job profiles, backgrounds, characteristics, and goals.

2. Understand your user's purpose and expectations when interacting with your AI. Consider the reason they use your AI product. For example, an empathetic tone might be necessary if your user uses the AI for customer service, while your AI can take a more authoritative tone for delivering information.

Consider Cultural Norms

When deploying an AI solution with a personality, you should consider the social and cultural values of the community within which it operates. This can affect the type of language your AI uses, whether to include small-talk responses, the amount of personal space, the tone of voice, gestures, non-verbal communications, the amount of eye contact, the speed of speech, and other culture-specific interactions. For instance, although a "thumbs-up" sign is commonly used to indicate approval, in some countries this gesture can be considered an insult.[83]

---

[83] Ethically Aligned Design: A Vision for Prioritizing Human Well-Being with Autonomous and Intelligent Systems. First ed., IEEE, 2019.

Designing Responses

Leveraging human-like characteristics within your AI product can be helpful, especially if product interactions rely on emulating human-to-human behaviors like having conversations, delegation, etc. Here are some considerations when designing responses for your AI persona.

Grammatical Person

The grammatical person is the distinction between first-person (I, me, we, us), second-person (you), and third-person (he, she, they) perspectives. Using the first person is useful in chat and voice interactions. Users can intuitively understand a conversational system since it mimics human interactions. However, using first person can sometimes set wrong expectations of near-perfect natural language understanding, which your AI might not be able to pull off. In many cases, like providing movie recommendations, it is better to use second-person responses like "You may like..." or third-person responses like "People also watched..."

Tone of Voice

What we say is the message. How we say is our voice.[84] When you go to the dentist, you expect a different tone than when you see your chartered accountant or your driving instructor. Like a person, your AI's voice should express personality in a particular way; its tone should adjust based on the context. For example, you would want to express happiness in a different tone than an error. Having the right tone is critical to setting the right expectations and ease of use. It shows users that you understand their expectations and goals when interacting with your AI assistant.

---

[84] Foster, Jonathan. "The Microsoft Voice, Part 1: You Had Me at Human." Medium, 2019, `https://medium.com/microsoft-design/microsoft-voice-principles-1-730f413190c1`.

An AI assistant focused on the healthcare industry may require some compassion, whereas an AI assistant for an accountant may require a more authoritative/professional tone, and an AI assistant for a real estate agency should have some excitement and enthusiasm.[85]

Strive for Inclusivity

In most cases, try to make your AI's personality as inclusive as possible. Be mindful of how the AI responds to users. While you may not be in the business of teaching users how to behave, it is good to establish certain morals for your AI's personality. Here are some considerations.

1. Consider your AI's gender or whether you should have one. By giving it a name, you are already creating an image of the persona. For example, Google Assistant is a digital helper that seems human without pretending to be one. That's part of the reason that Google's version doesn't have a human-ish name like Siri or Alexa.[86] Ascribing your AI a gender can sometimes perpetuate negative stereotypes and introduce bias. For example, an AI with a doctor's persona with a male name and a nurse's with a female name can contribute to harmful stereotypes.

2. Consider how you would respond to abusive language. Don't make a game of abusive language.

---

[85] "How to Design a Personality for Your AI Assistant." Medium, 2019, https://medium.com/aiautomation/how-to-design-a-personality-for-your-ai-assistant-4c43c320aa26.

[86] Eadicicco, Lisa. "Google Wants to Give Your Computer a Personality." time.com, 2017, https://time.com/4979262/google-wants-to-give-computer-personality/.

Don't ignore bad behavior. For example, if you say "Fuck you" to Apple's Siri, it denies responding to you by saying "I won't respond to that" in a firm, assertive tone.

3. When users display inappropriate behavior like asking for a sexual relationship with your AI, respond with a firm no. Don't shame people, but don't encourage, allow, or perpetuate bad behavior. You can acknowledge the request and say that you don't want to go there.

4. While it can be tempting to make your AI's personality fun and humorous, humor should only be applied selectively and in very small doses.[87] Humor is hard. Don't throw anyone under the bus, and consider if you are marginalizing anyone.

5. You will run into tricky situations when your users will say that they are sad or depressed, need help, or are suicidal. In such cases, your users expect a response. Your AI's ethics will guide the type of response you design.

Don't Leave the User Hanging

Ensure that your users have a path forward when interacting with your AI. You should be able to take any conversation to its logical conclusion, even if it means not having the proper response. Never leave users feeling confused about the next steps when they're given a response.

---

[87] Foster, Jonathan. "The Microsoft Voice, Part 2: Can We Talk?" Medium, 2019, https://medium.com/microsoft-design/the-microsoft-voice-part-2-can-we-talk-f83e3f502f7b.

Risks of Personification

While a human-like AI can feel more trustworthy, imbuing your AI with a personality comes with its own risks. The following are some risks you need to be mindful of:

1. We should think twice before allowing AI to take over interpersonal services. You need to ensure that your AI's behavior doesn't cross legal or ethical bounds. A human-like AI can appear to act as a trusted friend ready with sage or calming advice but might also be used to manipulate users. Should an AI system be used to nudge a user for the user's benefit or the organization building it?

2. When affective systems are deployed across cultures, they could adversely affect the cultural, social, or religious values of the community in which they interact.[88] Consider the cultural and societal implications of deploying your AI.

3. AI personas can perpetuate or contribute to negative stereotypes and gender or racial inequality, for example, suggesting that an engineer is male and a school teacher is female.

4. AI systems that appear human-like might engage in psychological manipulation of users without their consent. Ensure that users are aware of this and consent to such behavior. Provide them an option to opt out.

---

[88] Ethically Aligned Design: A Vision for Prioritizing Human Well-Being with Autonomous and Intelligent Systems. First ed., IEEE, 2019.

5. Privacy is a major concern. For example, ambient recordings from an Amazon Echo were submitted as evidence in an Arkansas murder trial, the first time data recorded by an artificial intelligence–powered gadget was used in a US courtroom.[89] Some AI systems are constantly listening and monitoring user input and behavior. Users should be informed of their data being captured explicitly and provided an easy way to opt out of using the system.

6. Anthropomorphized AI systems can have side effects such as interfering with the relationship dynamics between human partners and causing attachments between the user and the AI that are distinct from human partnership.[90]

A successful team of people is built on trust, so is a team of people and AI.

Summary

Building trust is a critical part of the user experience design process of AI products. This chapter discussed the importance of building trust, how you can build and reinforce trust with users, and pitfalls to avoid. Here are some important points:

---

[89] Eadicicco, Lisa. "Google Wants to Give Your Computer a Personality." time.com, 2017, https://time.com/4979262/google-wants-to-give-computer-personality/.

[90] Ethically Aligned Design: A Vision for Prioritizing Human Well-Being with Autonomous and Intelligent Systems. First ed., IEEE, 2019.

1. Your AI system will work alongside people and will make decisions that impact them. People and AI can work alongside each other as partners in an organization. To collaborate efficiently with your AI system, your stakeholders need to have the right level of trust. Building trust is a critical consideration when designing AI products.

2. Users can overtrust the AI when their trust exceeds the system's capabilities. They can distrust the system if they are not confident of the AI's performance. You need to calibrate user trust in the AI regularly.

3. Users need to be able to judge how much they should trust your AI's outputs, when it is appropriate to defer to AI, and when they need to make their own judgments. There are two key parts to building user trust for AI systems, namely, explainability and control.

4. **Explainability**

   Explainability means ensuring that users of your AI system understand how it works and how well it works. This allows product creators to set the right expectations and users to calibrate their trust in the AI's recommendations. While providing detailed explanations can be extremely complicated, we need to optimize our explanations for user understanding and clarity.

5. Different stakeholders will require different levels of explanation. Affected users and decision-makers often need simpler explanations, while regulators and internal stakeholders might not mind detailed or complex explanations.

6. The following are some guidelines to design better AI explanations:

    a. Make clear what the system can do.

    b. Make clear how well the system does its job.

    c. Set expectations for adaptation.

    d. Plan for calibrating trust.

    e. Be transparent.

    f. Optimize for understanding.

7. The following are the different types of AI explanations:

    a. Data use explanations

    b. Descriptions

    c. Confidence-based explanations

    d. Explaining through experimentation

8. Sometimes, it makes sense to provide no explanation when the explanations get in the way of a user's actions. Your users might already have a mental model of when and how that happens. It would also be wise to avoid explanations if they reveal private information or proprietary techniques.

9. You can consider evaluating your explanations through internal assessment with your team; qualitative methods like user interviews, surveys, etc.; or quantitative methods like usage metrics, product logs, etc. Quantitative methods are a good starting point to find the broad problem, but they need to be coupled with qualitative assessments.

10. **Control**

    Users should be able to second-guess the AI's predictions. You can do this by allowing users to edit data, choose the types of results, ignore recommendations, and correct mistakes through feedback. Users will trust your AI more if they feel in control of their relationship with it. Giving users some control over the algorithm makes them more likely feel the algorithm is superior and more likely continue to use the AI system in the future.

11. The following are some guidelines to design better control mechanisms:

    a. Balance the level of control and automation by considering the stakes of the situation and the time required for the AI and user to learn. You would need to provide stakeholders with greater control in high-stakes scenarios like surgery, healthcare, finance, or criminal justice.

    b. Hand off gracefully by returning control to the user when the AI fails or makes a mistake.

12. The following are the types of control mechanisms:

   a. **Data control**

   Your users should have the ability to view, access, edit, and share their data that the AI system uses in a private and secure manner. You should empower them through an explicit opt-in system and explainable information of how the AI stores and uses their data. You can allow users to select the data that the system uses and disable parts of the data they don't want to be used.

   b. **Control over AI output**

   Empower users to adapt your AI output to their needs, edit it, ignore it, or turn it off. You can provide a choice of results and allow users to correct the AI's mistakes through feedback. You can also make it easy for users to ignore the AI's result or allow them to dismiss AI outputs easily.

13. Apart from building trust through explanations and control mechanisms, you can also borrow trust from your customers, peers, service providers, reputable brands, or trustworthy organizations through social proof.

14. The following are some key trust-building opportunities for your AI:

   a. Onboarding

   b. User interactions

c. Loading states and updates

d. Settings and preferences

e. Error states

15. You need to be intentional about whether to give your AI a personality. While an AI that appears human-like might feel more trustworthy, your users might overtrust the system or expose sensitive information because they think they are talking to a human. An AI with a personality can also set unrealistic expectations about its capabilities.

16. It is generally not a good idea to imbue your AI with human-like characteristics, especially if it is meant to act as a tool like translating languages, recognizing objects from images, or calculating distances.

17. The following are some guidelines for designing an AI personality:

a. Don't pretend to be human.

b. Clearly communicate your AI's limits and capabilities.

c. Consider whom you are designing the personality for to help you make decisions about your AI's brand, tone of voice, and appropriateness within the target user's context.

d. Consider the social and cultural values of the community within which the AI operates.

e. Strive for inclusivity.

18.   Imbuing your AI with a personality can be
dangerous if used to manipulate people, engage
in psychological manipulation, or interfere with
relationship dynamics between humans. It can
sometimes perpetuate or contribute to negative
stereotypes and gender or racial inequality. Privacy
is also a major concern where your users might
overtrust the system or expose sensitive information
because they think they are talking to a human.

CHAPTER 5

Designing Feedback

Feedback is a critical component of designing AI products. This chapter will discuss the importance of feedback mechanisms, their types, and strategies for collecting and responding to feedback in the user experience. We will also discuss how you can give users control over their feedback information.

Like many people who worked in tech in 2020, I was lucky to have some free time at hand. So I decided to learn chess. While I knew the rules of the game, I wanted to learn it properly—tactics, strategies, openings, end games, etc. I signed up for an online class taught by Garry Kasparov (former world chess champion and the author of the book *Deep Thinking*). After spending a few months understanding chess theory, it was time to put my learnings to the test. I started practicing my game on an online chess app.

Chess has a rating system to estimate the strength of players. A grandmaster has a rating of more than 2500, while a novice has a rating of less than 1200. I began playing against a computer with a rating of 1800, and it decimated me. Then I lowered the rating settings to 800, which is when I started to win some games. Despite reducing the computer's rating, I cheated through some games by undoing moves or asking for hints. By doing this, I was correcting my mistakes and learning from them. Over time, my chess game improved. I learned two things from this experience: I could improve my game by playing lots of games, actively learning, and correcting my mistakes, and everyone else is much better at playing chess.

© Akshay Kore 2022
A. Kore, *Designing Human-Centric AI Experiences*,
https://doi.org/10.1007/978-1-4842-8088-1_5

The ability to learn is intelligent behavior. A lot of learning happens through feedback. In school, students learn when teachers correct their mistakes by providing feedback. Most students can understand a concept with enough repetition of output, correction, and input, that is, feedback loops. At work, I'm frequently part of UX design reviews. The goal of a design review is to provide feedback on a user experience to improve it. Restaurants collect feedback from customers to improve their experience. At the beginning of our lives, we have little understanding of the world around us, but we grow to learn a lot over time. We use our senses to take in data and learn via a combination of interacting with the world around us, being explicitly taught certain things by others, finding patterns over time, and, of course, lots of trial and error.[1] Feedback is information that comes back as a result of an action. Feedback loops are essential for any system that learns and improves over time.

Our natural environments are full of feedback loops. The processes of evolution and adaptation are, at their heart, a continuous feedback cycle. These feedback cycles are guided by reward systems. Evolution doesn't really care whether you have a brain or think interesting thoughts. Evolution considers you only as an agent, that is, something that acts.[2] For humans, characteristics such as intelligence, logical reasoning, planning, imagination, and creativity help in survival and progress. We are rewarded for these characteristics by our ability to adapt to and change environments to suit our needs, which improves our chances of propagating genes.

Our brains, too, have a reward system consisting of feedback loops mediated by dopamine. This internal signaling system connects positive and negative stimuli to behavior. We optimize for higher dopamine levels, which causes us to seek positive stimuli, such as sweet-tasting foods

---

[1] "IBM Design for AI." ibm.com, www.ibm.com/design/ai/.
[2] Russell, Stuart. *Human Compatible*. Allen Lane, an imprint of Penguin Books, 2019.

that increase dopamine levels. Seeking sugar-rich foods makes us avoid negative stimuli, such as hunger and pain, that decrease dopamine levels. Our brains come with built-in methods for learning so that our behavior becomes more effective at obtaining the dopamine reward over time. These methods also allow for delayed gratification so that we learn to desire things such as money that provide eventual reward rather than immediate benefits. One reason we understand the brain's reward system is because it resembles the method of reinforcement learning developed in AI.[3]

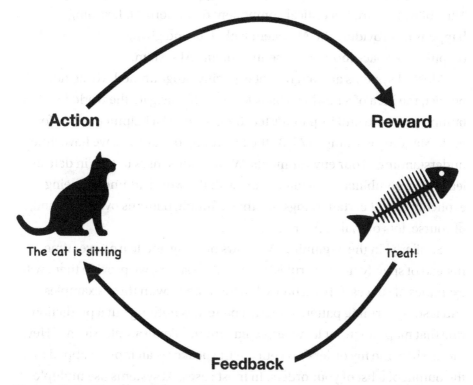

Action

The cat is sitting

Reward

Treat!

Feedback

Figure 5-1.   Feedback loop. *The cat gets a treat as a reward when it is sitting. Over time the cat would learn to sit whenever you show it a treat*

[3] Russell, Stuart. *Human Compatible*. Allen Lane, an imprint of Penguin Books, 2019.

Learning in AI happens by giving the system data and providing feedback on its outputs.

Feedback Loops in Artificial Intelligence

AI systems are not perfect. They are probabilistic and can sometimes make mistakes; they adapt and change over time. This change happens by correcting the AI's mistakes and identifying places where it can improve. The ability to learn is a critical component of modern AI. Learning happens by providing the AI system with data and giving feedback on its outputs. Feedback loops are essential for any AI system.

Most AI systems are taught, not explicitly programmed. While not perfect, the idea of a newborn baby learning to navigate the world by bumping around and its parents teaching it can be a helpful analogy for understanding learning in AI. At the beginning of our lives, we have little understanding of our environments. We use our senses to take in data and learn via a combination of interacting with the world around us, being explicitly taught certain things by others, finding patterns over time, and, of course, lots of trial and error.[4]

Similarly, in the beginning, AI knows nothing and learns over time. Instead of specifying exact rules, in most AI systems, we provide them with examples of successful outcomes. We train the AI with these examples, and it starts forming patterns. In AI, a pattern is nothing but a prediction rule that maps an input to an expected output.[5] For example, saying "Hey, Alexa, show me my orders" to your smart speaker is an input mapped to the output of a list of your orders. In most cases, AI systems use multiple inputs to predict an output. For example, an assistant suggesting you leave

---

[4] Kershaw, Nat, and C. J. Gronlund. "Introduction to Guidelines for Human-AI Interaction." Human-AI Interaction Guidelines, 2019, https://docs.microsoft.com/en-us/ai/guidelines-human-ai-interaction/.

[5] Polson, Nicholas G., and James Scott. *AIQ*. Bantam Press, 2018.

for the airport will need multiple inputs like your location, flight details from email, navigation data, traffic, weather conditions, etc. Different inputs can have different weights. This input-output mapping is called a model, and the process of finding a useful pattern is called training.

Training a model, that is, teaching an AI, starts with feeding it lots of examples. For instance, if you want to build an AI system that recognizes cats from images, you need to start showing lots of pictures of house cats to begin identifying patterns in them. Once it has built a pattern, you can evaluate its performance by showing it different images with and without cats in them. If it predicts a cat in an image that contains no cats, we give it negative feedback. If it predicts correctly, the model receives positive feedback. Based on the feedback it receives, the model calibrates itself. Over time as your AI system achieves an acceptable level of accuracy, that is, it predicts correctly much more often than it predicts incorrectly, you can deploy it to users. Now your users might test the product in different ways, some of which you might not have considered when training the AI. For example, users might show it an image of a tiger, which technically is a cat. Since you trained your AI on house cats, it outputs "Not a cat." This might lead to a mismatch in user expectations. To account for and capture this expectation, you can include feedback mechanisms in your product to improve it over time. The ability to give feedback when your AI makes mistakes can also help recalibrate user trust. It can help establish the mental model that your AI is learning and improving over time.

The ability to give feedback can help establish the mental model that your AI is learning and improving over time.

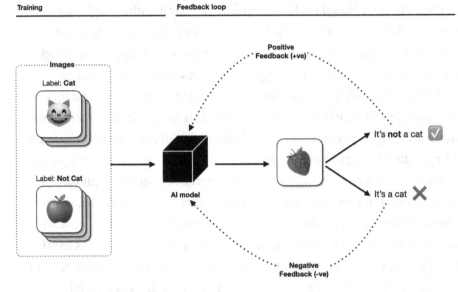

Figure 5-2. Cat detector. *In this example, we train the AI with labeled images containing cats and "no cats." Once it has built a pattern, you can evaluate its performance by showing it different images with and without cats in them. If it predicts a cat in an image that contains no cats, we give it negative feedback. If it predicts correctly, the model receives positive feedback*

Your AI system needs to be fast and accurate to be useful. While focusing on the speed of computations can be necessary for performance, speed alone is not enough for building a valuable system. Running a poorly designed algorithm on a faster computer doesn't make the algorithm better; it just means you get the wrong answer more quickly.[6] Along with faster computation, you need robust feedback mechanisms to build a useful AI product. Feedback is about improving the quality of results, and feedback is an important communication channel between your users, your product, and your team. Leveraging feedback is a

---

[6] Russell, Stuart. *Human Compatible*. Allen Lane, an imprint of Penguin Books, 2019.

powerful and scalable way to improve your technology and enhance the user experience.[7] Feedback mechanisms can help you improve your existing product and even discover new opportunities.

> *Feedback is an important communication channel between your users, your product, and your team.*

Types of Feedback

We tend to think of feedback as something users provide directly—through filling out surveys, sending feedback emails, etc. However, not all feedback is direct, and sometimes we can interpret it from user behavior and other sources. In AI products, users continuously teach the system to improve through direct and indirect responses.

The following are the three types of feedback:

1. Explicit feedback

2. Implicit feedback

3. Dual feedback

Explicit Feedback

Explicit feedback is when users intentionally give feedback to improve your AI system. This type of feedback is often qualitative, such as whether the AI suggestion was helpful or if the categorization and labeling were incorrect. Explicit feedback can take many forms like surveys, comments, thumbs-up or thumbs-down, open text fields, etc. By giving explicit feedback, users can feel more in control of the system. This type of feedback is extremely useful when you want unambiguous responses from your users about your AI's performance.

---

[7] pair.withgoogle.com, https://pair.withgoogle.com/.

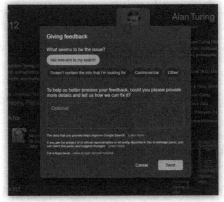

(a) Like/Dislike on Netflix

(b) Giving detailed feedback on Google search results

Figure 5-3.   Explicitfeedback. *(a) Like and dislike actions on Netflix titles. Source: Netflix website on desktop. (b) Giving detailed feedback on Google Search results. Source: Google Search website on desktop*

Using Explicit Feedback

You can use explicit feedback in the following ways:

1. When testing your AI in the early phases of product development, you can get explicit qualitative feedback from a diverse set of users from your target audience.[8] This can help in finding any glaring issues with your model or the training data.

2. After deploying your product to users, you can collect and review explicit feedback, categorize it into themes, and make changes accordingly. For example, feedback on a voice assistant not listening correctly can improve the model, while you can use feedback on its response to improve the logic.

---

[8] pair.withgoogle.com, https://pair.withgoogle.com/.

3. Sometimes explicit feedback can be fed directly into the AI model as a signal. For example, disliking a song in a music recommendation service changes all subsequent recommendations. If you're building such a feedback mechanism, ensure that the feedback data you receive can actually be used to improve the model.[9] Your team, your system, and your users should understand what the feedback means.

Guidelines for Incorporating Explicit Feedback

Explicit feedback allows users to give intentional reviews on your AI model. The following are some considerations when designing explicit feedback:

1. While asking users for feedback every time they use your AI system can be tempting, be mindful about intruding on the user's core workflow. It can be unpleasant if a popup shows up every time you click a title in a video streaming service. Ensure that you build appropriate and thoughtful feedback mechanisms at the right time in the user journey. For example, asking for feedback after the user has finished watching a movie is reasonable, but asking when they are in the middle of a scene isn't. Being mindful of intruding user workflows is especially important in high-stakes situations like driving or performing surgery.

---

[9] pair.withgoogle.com, https://pair.withgoogle.com/.

2. Whenever possible, explain the impact of user feedback on your AI and when it will show up. For example, liking a song in a music recommendation service can show a message "We will show more songs like this going forward." This indicates that the AI used the feedback to modify preferences, immediately impacting future recommendations.

3. Options for feedback responses should be mutually exclusive and collectively exhaustive.[10] For example, thumbs-up vs. thumbs-down is unambiguous and mutually exclusive and covers a full range of valuable opinions. It is actionable and understandable. On the other hand, a response like "Maybe" isn't actionable and can be ambiguous. In many cases, a neutral response can be equivalent to no response. If you need granular feedback, your options can match how users evaluate their experiences, for example, picking categories of music they are interested in.

4. Your explicit feedback mechanisms should be easy to understand. Your tone of voice should be appropriate for the kind of feedback you are requesting. Humor isn't appropriate when asking for something serious.

Note   In this chapter, we will mainly focus on explicit feedback mechanisms on the user interface.

---

[10] pair.withgoogle.com, https://pair.withgoogle.com/.

Implicit Feedback

Implicit feedback is data from user behavior and interactions from your product logs.[11] Implicit feedback is when we understand people's activities while using the product, which can be used to improve the AI model. Implicit feedback can be in the form of user interviews, customer service conversations, social media mentions, observations of user behavior, data from product logs, funnels in a user journey, or in-product metrics. This type of feedback is especially useful when validating hypotheses about user workflows. For example, suppose your team thinks that changing the ranking of results would improve the usage of an AI-based music recommendation service. In that case, you can validate this through usage metrics before and after the change, that is, implicit feedback.

---

[11] pair.withgoogle.com, https://pair.withgoogle.com/.

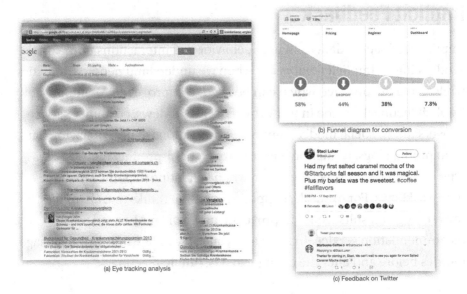

Figure 5-4. Implicit feedback. (a) Example of eye-tracking heatmap on a website. Source: www.researchgate.net/figure/Example-showing-what-a-heat-map-that-is-created-with-eyetracking-data-could-look-like-by_fig1_342783472. *(b) Funnel analysis for conversion on a website. Source:* www.hotjar.com/blog/funnel-analysis/. *(c) Positive feedback on Twitter. Source:* www.moengage.com/learn/marketing-automation-best-practices/

Using Implicit Feedback

You can use implicit feedback in the following ways:

1. When testing your AI in the early phases of product development, you can get implicit feedback through usability studies, observing user behavior with prototypes, etc. Implicit feedback can help find and fix any issues with the product and core workflows when interacting with the AI.

2. After deploying your AI system to users, you can
    monitor user behavior through in-product metrics
    and observing people interacting with your
    product. Reviewing product logs can help you find
    points of user confusion or frustration. This can
    help determine when and where your users need
    additional help or explanations to adjust their
    mental models. Depending on how frequently your
    AI is used, you can determine the precise moments
    to reinforce mental models.

3. You can learn from user behavior. Personalize the
    user's experience by learning from their actions
    over time.[12] For example, you can use previous
    inputs as implicit feedback to generate new
    recommendations or change the ranking of your
    suggestions.

Guidelines for Incorporating Implicit Feedback

Implicit feedback allows you to learn from user behavior and improve the
AI over time. The following are some considerations when incorporating
implicit feedback:

1. Let users know that you are collecting implicit
    feedback from data logs. Get user permission
    upfront and disclose it in your terms of service.
    Users should have the ability to see what data is
    being collected and how it is used in the product.
    Allow users to opt out of giving implicit feedback.

---

[12] Kershaw, Nat, and C. J. Gronlund. "Introduction to Guidelines for Human-AI
Interaction." Human-AI Interaction Guidelines, 2019, https://docs.microsoft.
com/en-us/ai/guidelines-human-ai-interaction/.

2. Whenever possible, explain how implicit feedback is used within your product interactions. For example, a section like "Because you watched movie X" in a movie recommendation service partially explains how the AI used their viewing behavior to generate a suggestion. It can also reinforce and recalibrate the mental models of your AI.

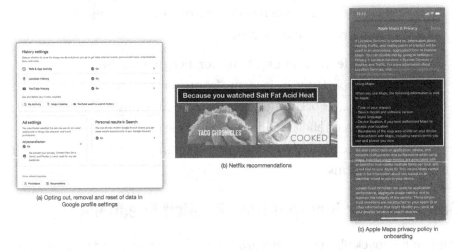

(a) Opting out, removal and reset of data in
Google profile settings

(b) Netflix recommendations

(c) Apple Maps privacy policy in
onboarding

Figure 5-5.   Incorporating implicit feedback. (a) Opting out, removing data, and resetting data on the Google account dashboard. Source: `https://myaccount.google.com/data-and-privacy`. *(b) A partial explanation of how implicit feedback was used in Netflix suggestions. The service recommends titles because the user watches Salt Fat Acid Heat. Source: Netflix website on desktop. (c) Apple Maps privacy policy has a clear and understandable explanation of data use, that is, how the system will collect implicit feedback. Source: Apple Maps app*

Dual Feedback

Dual feedback is a combination of implicit and explicit feedback. Sometimes you will receive both implicit and explicit feedback for a single feature. For example, users playing a song multiple times can be implicit feedback, and liking the same song can be explicit feedback. Both signals can help you tune your AI model. But, sometimes, there can be a mismatch between what users do and how they behave. For example, a user likes a particular song but skips it every time it is recommended. The user might have liked the song temporarily, and over time, their preferences might have changed. Feedback like this can be confusing because there isn't always a clear link between what a user does and what they want from the AI model.[13] In such cases, you need to consider using implicit and explicit feedback in tandem to tune your model. There is no easy solution for this, and this will depend on multiple factors like use case, type of user, frequency of use, etc. You will need to collaborate with your team to determine how to interpret certain actions and how much weight you should assign to different types of feedback.

---

[13] pair.withgoogle.com, https://pair.withgoogle.com/.

Spotify daily mix

*Figure 5-6.   **Dual feedback on Spotify's Daily Mix.** Daily Mix is an automatically generated playlist on Spotify based on your preferences. Playtime could be used as an implicit feedback signal; liking a song can be explicit positive feedback, while skipping is implicit negative feedback to the system*

Align Feedback to Improve the AI

Most physical products are static; they don't change over time. A bottle stays a bottle. It doesn't improve or change its properties based on what it contains. We are used to this static nature in many of our software systems. Most computer software follows strict logic—if this, then that. If a button is pressed, take a picture.

Contrary to this, AI models behave "softly"—they don't follow strict rules.[14] They are probabilistic and can be uncertain. A bald person's head and a duck egg can appear similar to an AI model—just a bunch of pink and brown pixels. If such incorrect results are presented as confident and discrete, it can confuse and erode user trust. Embrace this "softness" of your models, understand when and why they make specific mistakes, and provide appropriate mechanisms for user handoff and feedback.

AI products are dynamic; they change over time and adapt to their environments. For example, a music streaming service will adjust its recommendations based on user interactions with previous titles. Improvement in AI products happens through feedback, and your feedback mechanisms should help improve and tune your AI model. Before designing feedback mechanisms, let's understand how an AI model interprets feedback.

Reward Function

All AI models are guided by a reward function. Also known as an "objective function" or a "loss function," it is a mathematical formula (sometimes multiple formulas) that the AI model uses to determine correct vs. incorrect predictions.

The concept of the reward function is loosely based on a technique in psychology known as operant conditioning. Operant conditioning inspired an important machine-learning approach called reinforcement learning.[15] Operant conditioning has been used to train animals by giving them positive and negative reinforcement when performing a task. For example, if you want to train a dog to sit on command, whenever you say "Sit" and if the dog

---

[14] Webster, Barron. "Designing (and Learning from) a Teachable Machine—Library." Google Design, 2018, https://design.google/library/designing-and-learning-teachable-machine/.

[15] Mitchell, Melanie. *Artificial Intelligence*. First ed., Farrar, Straus and Giroux, 2019.

sits down, you offer it a reward in the form of a treat. Over time the dog learns to sit when it hears the words "Sit down." In this case, you saying the words "Sit down" is the input, the dog hearing the words and sitting down is the action, the act of giving a treat (or not giving one) is feedback, and the treat itself is the reward. Similarly, when you provide feedback to an AI model, it calibrates or tunes the reward function. A well-tuned reward function is right more times than it is wrong. This results in an improvement in the AI model.

Reward behavior I like and ignore (or punish) behavior I don't.

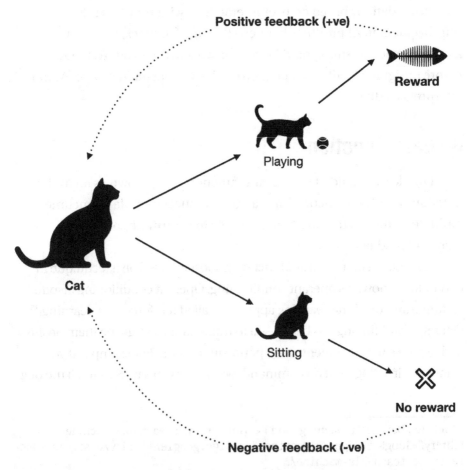

Figure 5-7. *Training a cat using positive and negative reinforcement*

Collaborate with Your Team

The performance of your AI product is guided by its reward function. The feedback mechanisms you incorporate should help calibrate the reward function and thereby tune your AI model. But remember that the process of designing feedback mechanisms should be collaborative across product functions. Extend your design team to include product managers, engineering and machine learning teams, and UX counterparts. Sometimes, you can also include customer success teams and certain users when designing feedback mechanisms.

The following are some considerations when collaborating with your team:

1. Think about possible outcomes of providing feedback and bounce off ideas with your team members. This will help you identify pitfalls and points of confusion. Sometimes, feedback might not be helpful or align with tuning your model. For example, in a workout recommendation service, users might want to give feedback so that the model recommends more videos from a particular instructor, but the feedback is interpreted to show more exercises of the same type like yoga or HIIT. This can lead to a mismatch in expectations and erode user trust.

2. Discuss the tradeoffs of collecting or not collecting feedback. Strong explicit signals are great for tuning your model, but be thoughtful about how you interpret those signals. Sometimes, looking at interactions over a longer period of time can help you distill more accurate patterns of behavior and intent.[16]

---

[16] pair.withgoogle.com, https://pair.withgoogle.com/.

3. Discuss opportunities for collecting implicit and explicit feedback across the user journey. For example, clicks on food images can be interpreted as implicit feedback in a restaurant recommendation app, while star ratings are explicit feedback.

4. Consider the context and stakes of the situation. In busy or high-stakes scenarios like driving or performing surgery, users might not be able to provide explicit feedback.

5. AI models can age and become less useful over time. In such cases, feedback on earlier results might become irrelevant. Discuss how often the underlying data will be updated with your machine learning teams and consider if you should communicate this change to users.

6. Don't dictate exactly how to collect feedback, and don't be prescriptive. Consider designing feedback as a collaborative exercise with different points of view.

7. Plan ahead in the product cycle for long-term research and short-term feedback. Allocate enough time to evaluate the performance of your AI models through qualitative and quantitative measures. This can help you validate the user experience, your AI's performance, and any other hypotheses over different time spans.

8. Think about how you can get real user feedback through observation and user interviews to improve the AI model. Understand how mental models of your system evolve over time. Communicate learnings regularly with your team.

9. Make a list of as many events and opportunities
    for collecting feedback that can improve your AI
    model. Cast a wide net—app Store reviews, Twitter,
    email, call centers, push notifications, etc. Then,
    systematically ask what your users and data are
    telling you about your product's experience.[17] Here
    are some considerations when making this list:

    a. What is the user experience that triggered
        this opportunity? For example, errors,
        recommendations, login, etc.

    b. Is the feedback implicit or explicit?

    c. What are users providing feedback on? For example,
        slow loading time, accidental clicks, ranking issues,
        inappropriate recommendations, etc.

Feedback enables users to teach the AI.

Collecting Feedback

Feedback mechanisms in your AI system should help you improve the
AI model. The process of collecting feedback is the process of enabling
users to teach the AI. Feedback may be fed directly into the AI model, or
your team can collate feedback, analyze it, and make changes at a later
time. However, your users might not be inclined to give feedback all the
time. When collecting feedback, ensure that feedback mechanisms don't
get in the way of core user workflows or stop users from doing their job. A
restaurant app asking for feedback every time you select an item will be a
frustrating experience, while collecting feedback after the food is delivered

---

[17] pair.withgoogle.com, `https://pair.withgoogle.com/`.

is reasonable. The following are some considerations when designing feedback mechanisms:

1. Consider the stakes of the situation.

2. Make it easy to provide feedback.

3. Explain how feedback will be used.

4. Consider user motivations.

Giving feedback should be optional and should not intrude on the user's core workflow.

Consider the Stakes of the Situation

Consider the type of situation in which your AI will be used. The stakes of your situation will determine the level of intrusion of your feedback mechanisms. A popup in the middle of a workflow is highly intrusive, while interface elements like flags, thumbs-up or thumbs-down, etc. are relatively less intrusive. It would not be advisable to show highly invasive feedback mechanisms when the stakes are high, like performing surgery, driving a car, or making a large transaction. In such cases, it is better to include less intrusive strategies. Sometimes, you might want to avoid collecting explicit feedback altogether and gather feedback implicitly.

You can consider including intrusive feedback mechanisms in low-stakes scenarios like providing music recommendations or choosing restaurants. Even in such cases, it is advisable to use feedback mechanisms that are not too intrusive, and you can always collect feedback implicitly. However, there can be cases where getting user feedback is absolutely essential. For instance, in case of the failure of a business-critical application like messaging or payments, where the message or payment failed to send, collecting feedback can help you find the root cause of failure. However, providing feedback should be optional in most cases and should not intrude on the user's core workflow.

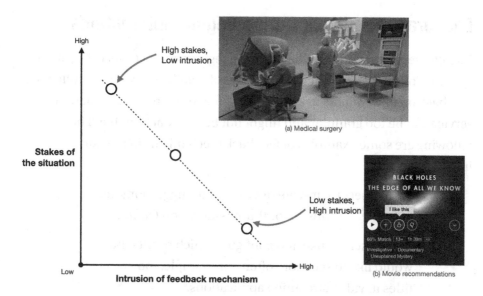

Figure 5-8. Intrusion of feedback vs. stakes of the situation.
*(a) Performing surgery is a high-stakes situation. It is better to
include less intrusive mechanisms. Sometimes, you might want to
avoid altogether collecting explicit feedback and gather feedback
implicitly. Source:* `https://surgical-solutions.com/blog/what-`
`is-robotic-surgery-and-how-does-it-work/`*. (b) Recommending
movies is a low-stakes scenario. You can consider including intrusive
feedback mechanisms. Source: Netflix website on desktop*

Make It Easy to Provide Feedback

While a survey form is commonly used to collect feedback, you can go
beyond forms and input fields when designing feedback mechanisms
for your AI. Whenever possible, collect feedback implicitly. Create
opportunities for providing feedback when the AI makes mistakes and
even when it provides a correct prediction. Encourage users and make it
easy for them to provide feedback.

Encourage Feedback During Regular Interactions

Enable users to provide feedback during regular interactions with your AI system. Collect feedback at the right level of detail. For example, collecting feedback on instruments used in a song for a music recommendation service can be too granular and might not be easily actionable. The following are some examples of feedback mechanisms in regular interactions:

1. Allow users to up-vote or down-vote suggestions, for example, thumbs-up or thumbs-down on Quora.

2. Allow users to like and configure which results they would like to see more often, for example, movie titles in video streaming applications.

3. Allow users to provide ratings, for example, star ratings on restaurants in a food delivery application.

4. Change the ranking of results, for example, shuffling the order of songs in an AI-generated playlist.

5. Progressively disclose survey questions throughout the user experience when appropriate. For example, the Apple Watch's handwashing timer asks for feedback if you finish cleaning under the 20-second limit, or Netflix asks for feedback after you've watched a title.

6. Provide utility in the form of saving, bookmarking items, or creating personalized lists. Actions on these can also be interpreted as feedback for your AI. For example, bookmarking news articles results

in suggesting similar articles on a news service, highlighting points of interest in an audiobook results in recommending similar titles, or tagging photos in a social media application is sometimes used for training a facial recognition model. Be mindful of any privacy implications; you should explicitly communicate this data collection and allow users to opt out.

(a) Google News   (b) Netflix thumbs up/down   (c) Apple Watch hand washing timer

Figure 5-9.   Encourage feedback during regular interactions. (a) Users can easily give feedback on news article recommendations on Google News. Source: Google News on Chrome. (b) Netflix makes it easy to like, dislike, or save a title to your list. Source: Netflix website on desktop. (c) Apple Watch can automatically detect when a user is washing their hands to start a 20-second timer. The handwashing timer app asks users for feedback if they stop washing their hands before 20 seconds are up. Source: www.myhealthyapple.com/how-to-turn-off-apple-watch-handwashing-reminders-and-features/

Allow Correction When the AI Makes Mistakes

Errors are opportunities for feedback. Make it easy for users to correct your AI's mistakes by editing, refining, overriding, or undoing the system's results. You can use instances of error correction as implicit or explicit feedback. The following are some examples:

1. Allowing users to edit automatically generated results, for example, editing automatic tags of people on photos in a social media application

2. Allowing users to undo inferences made by the system, for example, manually reverting autocorrected text

3. Allowing users to remove or dismiss suggestions, for example, removing or dismissing irrelevant results in a video streaming service

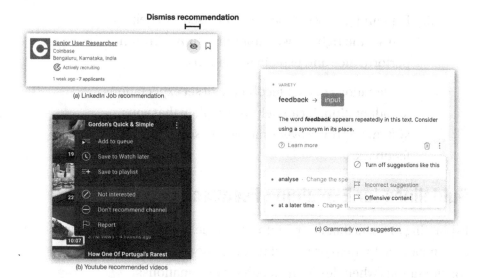

Figure 5-10. *Allow correction when the AI makes mistakes. (a) Hiding and dismissing irrelevant job opportunities on LinkedIn. Source: www.linkedin.com/. (b) Users can flag a recommended video on YouTube that they are not interested in. Source: www.youtube.com/. (c) Grammarly allows users to flag incorrect suggestions. Source: Grammarly plugin on desktop*

Explain How Feedback Will Be Used

Users are more likely to provide feedback if they understand how it is used and how it benefits them. You can provide explanations before asking for feedback, for example, explaining how their ratings will be used in a restaurant recommendation service. You can also give explanations after users provide feedback; for example, liking a song in a music streaming service shows the following message: "We will show you more songs like this." The following are some benefits of explaining how you will use AI feedback:

1. Explaining the usage of feedback allows you to reinforce and strengthen the mental models of your AI system. It can also help you adjust user expectations.

2. Explaining how feedback will be used can help you set the right expectations of your AI's current performance and explain limitations.

3. Explanations of user feedback can set expectations for adaptation. It can help you explain that your system improves over time when users provide feedback.

Guidelines for Explaining Feedback Use

Explaining to users that their feedback will help improve the system can encourage them to provide it readily. The following are some considerations when designing feedback explanations:

1. **Explain how the AI will improve.**

   Whenever possible, users will avoid giving feedback. If the benefit of providing feedback isn't clear and specific, users may not understand why they should give feedback. Explain how the feedback will be used while asking for feedback or after feedback is provided. Be clear about what information will help your AI learn and how it will improve the user experience. Connect feedback mechanisms with changes in the user experience. For example, you can tie user benefits with feedback by using messaging like "You can improve your experience by providing feedback on suggestions." When users like your AI's suggestion, consider responses like "We will show you more content like this."

2. **Explain when will the AI improve.**

    Sometimes AI feedback is directly fed into
    the system. For example, a like or dislike on a
    personalized music or video streaming service can
    immediately change future suggestions. However,
    in most cases, even with the best data and feedback,
    AI model improvements are almost never possible
    to implement immediately.[18] Your team might need
    to wait and collect lots of feedback before making
    changes to the model and deploy it at a later time
    as an update. You can design your messaging to set
    clear expectations of when user feedback will take
    effect. In cases where feedback is used immediately,
    you can use messaging like "We will start showing
    more content like this from now on." When feedback
    is used later, you can thank users for providing
    feedback and explain that your team will use it to
    improve the system. You may or may not decide
    to provide a definite time when such feedback will
    take effect.

3. **Set expectations for improvement.**

    When collecting feedback, explain how teaching the
    AI helps it improve over time and benefits the user.
    Clarify how their feedback improves the system. For
    example, explain that providing a difficulty rating to
    an exercise in a workout app can help the AI suggest
    more relevant workouts.

---

[18] pair.withgoogle.com, https://pair.withgoogle.com/.

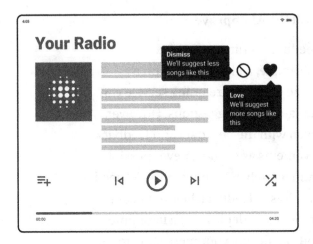

A personalised playlist based on your listening habits. Improves as you listen more.

Figure 5-11. Music recommendation. The messaging highlights how the AI improves over time and how the user can help the system by liking or disliking songs, that is, giving explicit feedback

Users are more likely to give feedback if it gives them direct or indirect benefits.

Consider User Motivations

Star ratings, Likert scales, survey forms, comment boxes, or thumbs-up/thumbs-down buttons are not the only mechanisms for collecting feedback for your AI. While explaining how feedback is used can nudge some users to give more of it, it needs to be valuable for users to take the time to provide feedback. Understand why people give feedback. The value of providing feedback is tied to motivation, and users are more likely to give feedback if it gives them direct or indirect benefits.

Consider the following user motivations when designing feedback mechanisms:

1. Reward

2. Utility

3. Altruism

4. Self-expression

Reward

A reward is the simplest form of incentive. Some users may be more inclined to provide feedback if they receive material or symbolic rewards in return. You can tie your feedback mechanisms with the product's reward systems. The following are the types of rewards that you can give users in exchange for feedback.

Symbolic Rewards

Symbolic rewards can be in the form of status updates, levels, or virtual badges in your product. For example, an app that encourages users to exercise regularly can give them badges to keep them motivated. The following are some advantages and disadvantages of symbolic rewards:

- **Advantages**

    - They have a very low to no cost.

    - They can be freely distributed without significant implications.

- **Disadvantages**

    - Your users need to care about the rewards for them to be motivated.

    - Symbolic rewards can inhibit intrinsic motivation.

  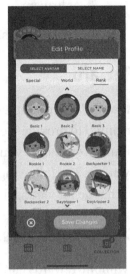

(a) Awards on Apple fitness   (b) Reading insights on Amazon's Kindle app   (c) Badges in the game Two Dots

Figure 5-12. Symbolic rewards. (a) Award badges on Apple Fitness. Source: Apple Fitness app on iOS. (b) Reading insights on Amazon's Kindle app. Source: Amazon Kindle mobile app. (c) Progress badges in the game Two Dots. Source: Two Dots game by PlayDots

Material Rewards

Material rewards can be in the form of cash payments, coupons, discounts, or even giving free products. For example, many stock trading platforms offer one free share to new users. Along with providing an incentive to sign up, this can teach them how to use the product and collect feedback by analyzing behavior, noticing drop-offs, etc. Many machine learning labeling services offer their partners money in exchange for training the AI and giving feedback. The following are some advantages and disadvantages of material rewards:

- **Advantages**

  - Material rewards are easy to understand.

  - It is a straightforward solution to increase feedback.

  - It can lead to a large volume of feedback.

- **Disadvantages**

  - They are costly to run for long durations.

  - They may inhibit or devalue intrinsic motivation.

  - There is a high risk of drop-off as soon as the reward stops.

  - You might not be able to collect feedback from the right users.

  - The quality of your feedback may reduce.

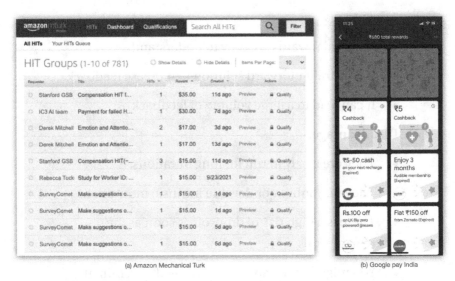

(a) Amazon Mechanical Turk   (b) Google pay India

Figure 5-13. Material rewards. (a) Amazon Mechanical Turk is a crowdsourcing marketplace that makes it easier for individuals and businesses to outsource simple tasks like tagging images, transcription, etc. in exchange for money. Source: www.mturk.com/. (b) Google Pay India app gives users rewards like discounts, coupons, and even cash for using their service. Source: Google Pay India app

Social Rewards

Also known as status in society, social wealth or social capital can be a powerful motivator; this can be in the form of social proof, belonging to a group (or cult), projecting a self-image, or reputation as a leader or an expert. Social rewards are based on perceptions within a society. Symbolic rewards can sometimes be converted into social capital by sharing with a community, social networks, or leaderboards. For example, a verified tick on Twitter is a type of social capital that establishes you as someone with considerable influence. It encourages users to engage more with the service by creating an aspiration. The following are some advantages and disadvantages of social rewards:

- **Advantages**

  - They have very low to no cost.

  - Social rewards can create network effects in a community and thereby encourage more feedback.

  - They provide powerful intrinsic motivations.

- **Disadvantages**

  - Your users need to care about the rewards for them to be motivated.

  - You need existing network effects for them to be valuable.

  - They can lose their value if they are freely distributed or are easily attainable.

(a) Twitter verified account

(b) Views, likes and subscriber count on Youtube

(c) League of Legends leaderboard

Figure 5-14.   Social rewards. *(a) Verified badge on Twitter. Source:* https://twitter.com/Twitter. *(b) Views, likes, and subscriber count on YouTube. Source:* www.youtube.com/. *(c) Online leaderboard for the game League of Legends. Source:* www.leagueofgraphs.com/rankings/summoners

Utility

Users can be motivated to give feedback if the feedback mechanism is also useful. Utility-based motivations can include features like bookmarking, creating lists, saving results, tracking progress, or quantified self-experiences like saving current health status. You can tie usage of helpful features with feedback mechanisms. For example, bookmarking a news article can be used as a signal to recommend similar articles. The following are some advantages and disadvantages of utility-based motivations:

- **Advantages**

  - No network effects are necessary.

  - Users are intrinsically motivated to use the feature and thereby give feedback.

  - There is a direct relationship between helping users and collecting feedback.

- **Disadvantages**

  - Sometimes, you cannot build network effects due to privacy constraints.

  - Highly dependent on the quality and usage of the utility feature.

(a) Bookmarking an article in Google News

(b) Tracking workouts on Apple fitness

(c) Liking, hiding and adding songs to a playlist on Spotify

Figure 5-15. *Utility.* *(a) Bookmarking an article on Google News. Source: Google News on Chrome. (b) Automatically tracking workouts on Apple Fitness. Source: Apple Fitness app on iOS. (c) Liking, hiding, and adding songs to a playlist on Spotify. Source: Spotify app on iOS*

Altruism

Sometimes people are just nice; they want to help you improve your product by leaving a product review or agreeing or disagreeing with a comment on your product. Sometimes they want to help other people by creating a detailed video or post-highlighting the pros and cons of your product. The following are some advantages and disadvantages of altruistic motivations:

- **Advantages**

  - Users are intrinsically motivated to help.

  - Potential for more honest feedback.

- **Disadvantages**

    - In most cases, your product needs to be established.

    - Altruism varies across cultures and groups.

    - The quantity of feedback may be low.

    - The quality of subsequent feedback may be biased if a reviewer has high social capital.

    - Feedback on social media may be more extreme than ground reality.

(a) Restaurant review on Zomato

(b) Upvote, downvote and comments on Quora

(c) Product review on Amazon

Figure 5-16.   Altruism. (a) Restaurant review on Zomato. Source: www.zomato.com/. (b) Up-votes, down-votes, and comments on Quora. Source: www.quora.com/. (c) Product reviews on Amazon. Source: www.amazon.in/

Self-Expression

Many people get intrinsic fulfillment by expressing themselves. They get enjoyment from community participation. Feedback derived from self-expression can be in the form of comments on forums, venting about the product on social media, customer care complaints, etc. The following are some advantages and disadvantages of self-expression-based motivations:

- **Advantages**

  - Users are intrinsically motivated to help.

  - Potential for more honest feedback.

  - Potential for getting trustworthy and critical feedback.

- **Disadvantages**

  - Feedback may be more extreme than ground reality.

  - Feedback can be more biased since it is dependent on individual preferences.

  - Self-expression varies across cultures and groups.

(a) Review on Reddit

(b) Complaint on Twitter

(c) Tutorial on Youtube

Figure 5-17. Self-expression. (a) A user's self-initiated review of a product on Reddit. Source: www.reddit.com/. (b) Complaint on Twitter. Source: www.awesomeinventions.com/hilarious-customer-complaints-company-responses/. (c) Tutorial on driving a manual car on YouTube. Source: www.youtube.com/

By considering user motivations, you can incorporate more effective and varied feedback mechanisms to improve your AI. In most cases, feedback for your AI would be a combination of explicit and implicit feedback mechanisms, including reward-, utility-, altruism-, and self-expression-based motivations.

Responding to Feedback

When users provide feedback to your AI, they need to know that the feedback was received. Your interface should confidently respond to user feedback. A weak response or, worse, no response can erode user trust. Imagine clicking a remove suggestion button, and nothing happens, or writing and submitting a feedback form does not confirm that the feedback was sent. You should provide an immediate confirmation when users give feedback. Sometimes, you might want to offer additional summaries of feedback at a later time.

When users give feedback, they need to know that it was received.

On-the-Spot Response

When users give feedback, they should immediately know whether the system received it and when the AI or your team will act on it. Respond to feedback on the spot. The following are some guidelines for designing on-the-spot responses.

Connect Feedback to Changes in the User Experience

Connect feedback to changes in the user experience. Acknowledging that you've received the user's feedback can build trust. You should also let users know what the system will do next or how their input influences the AI.[19] For example, dismissing or disliking a suggestion can remove it from a list of recommendations.

---

[19] pair.withgoogle.com, https://pair.withgoogle.com/.

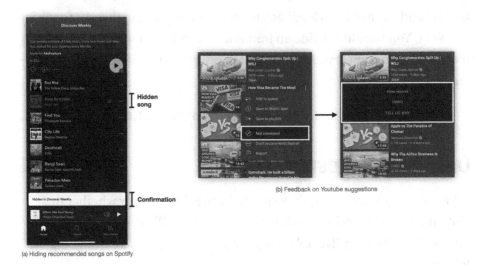

(a) Hiding recommended songs on Spotify

(b) Feedback on Youtube suggestions

Figure 5-18. Connect feedback to changes in the user experience.
(a) When a user removes a song from a personalized playlist on
Spotify, the interface indicates that the song is hidden by changing its
status and showing a confirmation. Source: Spotify app. (b) Marking
a video as "Not interested" on YouTube recommendations removes it
from the view. Source: www.youtube.com/

Clarify Timing and Scope

Whenever possible, your response should clarify when the feedback will
be acted on and what might change. Timing explains when user feedback
will impact the system; scope clarifies what would change and how
much. However, it is not always possible to communicate when and how
user feedback will affect the AI. In such cases, you can provide a generic
message like "Thank you for your feedback." But if you do communicate
timing and scope, make sure you deliver on the promise. For feedback that
is immediately fed into the AI system, you can show the impact right away.
If you think detailed clarifications can confuse users, you can make your
explanations ambiguous.

The following are some examples of messaging that clarify timing and scope in a music streaming service:

1. "Thank you for your feedback."

   • Timing: Not specified. Generic response

   • Scope: Not specified

2. "Thank you! Your feedback helps us improve future suggestions."

   • Timing: Broadly in the future

   • Scope: Suggestions of all users

3. "Thank you! Your feedback helps us improve your future suggestions."

   • Timing: Broadly in the future

   • Scope: Your suggestions

4. "Thank you! We won't recommend this artist to you."

   • Timing: Now, but ambiguous

   • Scope: Your suggestions from this artist

5. "Thank you! We'll suggest more songs like this from now on."

   • Timing: Now

   • Scope: Your suggestions, but ambiguous

6. "Thank you! We've updated your preferences. Take a look."

   • Timing: Now

   • Scope: Your preferences with a view of the change

Set expectations for adaptation

AI products change over time by adapting to user preferences. By providing feedback, your users are helping the AI to improve. Explain what your system can do, how well it performs, and how it improves over time. When users provide feedback, explain how teaching the AI benefits them. Setting expectations for adaptation can help you build and reinforce mental models of your AI. It can help you build and recalibrate user trust.

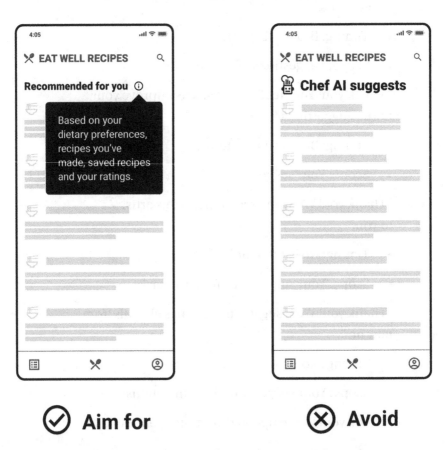

Figure 5-19. Recipe recommender: Set expectations for adaptation. (Left) The tooltip explains how the system generated the recipe recommendations based on the app's user activity. (Right) The recommendation on the right is ambiguous

Limit Disruptive Changes

Limit disruptive changes to the user experience after users provide feedback. Update and adapt cautiously.[20] If you are modifying results, do it gradually. Only make changes that are relevant to the input; don't change all results unless specified. For example, if the user removes one article from the suggestions in a news application, don't change all suggestions. Unless users explicitly ask for it, don't refresh all suggestions. It is a jarring experience when items disappear without explanations when they were previously available.

Long-Term Response

Sometimes, along with providing immediate responses, you might want to give a summary of feedback and changes over a period of time. For example, a service that tracks your spending and suggests investment opportunities might send a weekly report. In 2020, Spotify created an experience called "Spotify Wrapped" to create a narrative summary of users' listening habits for the year. A long-term response can be in the form of periodic reports, summaries, developer documentation, system update notifications, versioning, onboarding new features, etc. Inform users when the AI system adds to or updates its capabilities.[21] Additionally, you can let users know what feedback resulted in the change.

---

[20] Kershaw, Nat, and C. J. Gronlund. "Introduction to Guidelines for Human-AI Interaction." Human-AI Interaction Guidelines, 2019, `https://docs.microsoft.com/en-us/ai/guidelines-human-ai-interaction/`.

[21] Kershaw, Nat, and C. J. Gronlund. "Introduction to Guidelines for Human-AI Interaction." Human-AI Interaction Guidelines, 2019, `https://docs.microsoft.com/en-us/ai/guidelines-human-ai-interaction/`.

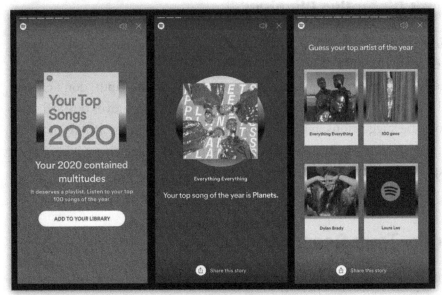

Spotify Wrapped

Figure 5-20. Spotify Wrapped. Spotify created an experience called "Spotify Wrapped" to create a narrative summary of a user's listening habits in a year. Source: www.techradar.com/in/news/spotify-wrapped-2020-launches-to-remind-music-lovers-of-a-year-wed-rather-forget

Control

Give people control over the feedback they give your AI. If people feel more in control, they are more likely to trust your system. In most cases, giving explicit feedback should be optional. Allow users to see and edit feedback already given by them. The following are some considerations when giving users control over feedback.

If people feel more in control, they are more likely to trust your AI.

Editability

Your user preferences can change. Sometimes users might give incorrect feedback or make mistakes while giving it. You should allow users to edit, update, or undo feedback given and change preferences at a later time. You can do this as a part of the feedback response or allow users to configure feedback in your product's settings.

(a) Undo feedback on Youtube suggestions

(b) Preference editing on Flipboard

Figure 5-21. *Editability. (a) Undo feedback on YouTube suggestions. Source: www.youtube.com/. (b) Preference editing on Flipboard. Source: Flipboard app on iOS*

Removal and Reset

Sometimes users might not want the AI to access the feedback they provided earlier. Allow users to remove and reset past feedback and preferences. People change; a user can start liking a genre of music that they disliked in the past. The ability to remove and reset is beneficial when most of your AI's results are irrelevant or not useful. Sometimes, the best way then is to start from scratch. For example, a user accesses a video streaming service after some years. By then, their preferences would have changed, and the current results might not be relevant. While it is possible to calibrate results by giving additional feedback over time, this process can be tedious. The best course of action is to reset and start from scratch.

Opting Out

In most cases, giving feedback should be optional. Allow users to opt out of giving feedback. In particular, you should allow users to opt out of certain aspects of sharing implicit feedback—like having their behavior logged—and this should be included in your terms of service.[22] Respect your user's decision to opt out. If collecting feedback is absolutely necessary for your AI to work, clarify that they might not be able to use the product. If possible, provide a manual, non-AI workflow. Remember that user preferences can change and they might want to provide feedback at a later point to use the AI. Along with opt-out, also include an opt-in mechanism.

> *Giving feedback should be optional. Along with opt-out, also include an opt-in mechanism.*

---

[22] pair.withgoogle.com, https://pair.withgoogle.com/.

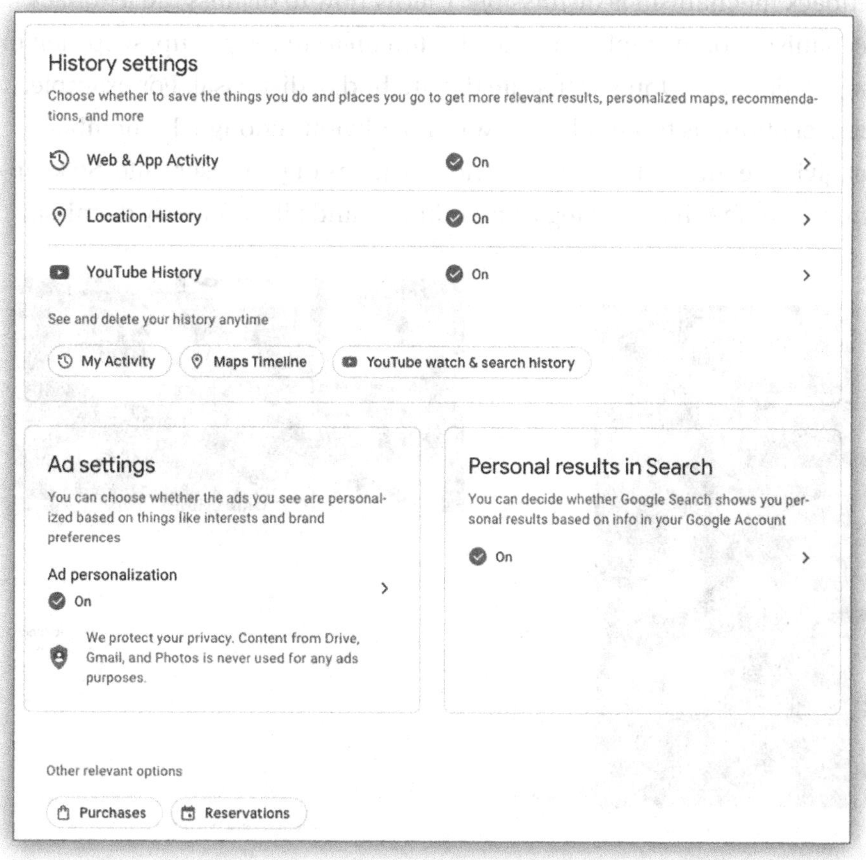

Editing preferences, opting out, removal and reset of data in Google profile settings

Figure 5-22.   Editing preferences, opting out, removing and resetting data on the Google account dashboard. Source: `https://myaccount.google.com/data-and-privacy`

Make It Easy to Ignore and Dismiss

Most times, users don't want to give feedback. Your feedback mechanisms should not get in the way of the user's core workflow. A feedback popup that obstructs your workflow can be frustrating. Make it easy to ignore giving feedback. For example, a thumbs-up or thumbs-down button is unobtrusive. If you are explicitly asking for feedback, ensure your feedback mechanism is dismissible. Clarify how to dismiss the feedback mechanism, for example, close the button on a survey popup, swipe left on a suggestion, etc. Don't obfuscate the method of dismissal. For example, a cancel button is too small to view or not obvious enough. Remember that giving feedback is most likely not the focus of your user's life. So keep requests for feedback strategic and minimal and allow for easy dismissal.

(b) Like/Dislike on Netflix

(a) Knowledge panel on Google Search

(c) Job recommendations on LinkedIn

Figure 5-23. Make it easy to ignore and dismiss feedback mechanisms. *(a) The feedback mechanism occupies a small real estate on the knowledge panel on Google Search. Source:* `www.google.com/`*. (b) Liking and disliking movie titles does not obstruct the user's main workflow of selecting and watching a movie on Netflix. Source: Netflix website on desktop. (c) Feedback mechanisms like dismissing a job recommendation or bookmarking have less weight on the UI. Source:* `www.linkedin.com/jobs/`

> *Remember that giving feedback is most likely not the focus of your user's life. So keep requests for feedback strategic and minimal and allow for easy dismissal.*

Transparency

Users aren't always aware that their inputs and behaviors are being used as feedback. Get user permission upfront. Clearly communicate what feedback is being collected, what data is personal, what is shared, and how it is used. Communicating this information in a clear, understandable manner is critical. Giving users access to a usage log with a million entries is not understandable. Providing this transparency and explaining it in a human-understandable way can help your users calibrate the right level of trust in your AI.

(a) Google Assistant Onboarding

(b) Apple Maps privacy policy in onboarding

Figure 5-24. Transparency. (a) Onboarding for Google Assistant summarizes how data is collected, shared, and used. Source: Google Assistant app. (b) Apple Maps privacy policy shown during onboarding has a clear and understandable explanation of data use. Source: Apple Maps app

Human-AI Collaboration

Instead of thinking of an AI system as something that operates in a silo, people and AI can be symbiotic partners in a system. Teams of humans and AI are collectively more intelligent, adaptable, and resilient. AI can give people superpowers, while people can teach and improve it through feedback. AI systems are continuously learning. Learning happens through consuming data, forming models, and correcting mistakes, that is, feedback loops. You can tightly integrate your AI feedback mechanisms with your processes, workflows, and strategies. Sometimes you might want to rethink processes entirely to build these symbiotic relationships between people and AI.

AI can give people superpowers, while people can teach and improve it through feedback.

Figure 5-25. *Human-AI collaboration. Source: Photo by ThisisEngineering RAEng on Unsplash*

Summary

Designing robust feedback mechanisms that help you improve your AI is a vital component of building great AI experiences. In this chapter, we discussed various strategies for collecting and responding to feedback and giving users control over their feedback information. Here are some important points:

1. The ability to learn is a critical component of modern AI. Learning in an AI system happens by providing it with data and giving feedback on its outputs. Feedback loops are essential for any AI system.

2. We tend to think of feedback as something users provide directly through filling out surveys, sending feedback emails, etc. However, not all feedback is direct. Sometimes we can interpret feedback from user behavior and other sources. In AI products, users are continuously teaching the system to improve through direct and indirect responses. The following are the three types of feedback:

    a. Explicit feedback

    b. Implicit feedback

    c. Dual feedback

3. Explicit feedback is when users intentionally give feedback to improve your AI system. Explicit feedback can take many forms like surveys, comments, thumbs-up or thumbs-down, open text fields, etc.

4. Implicit feedback is data from user behavior and interactions from your product metrics. Implicit feedback can be in the form of user interviews, customer service conversations, social media mentions, observations of user behavior, data from product logs, funnels in a user journey, or in-product metrics.

5. Dual feedback is a combination of implicit and explicit feedback. Sometimes you will receive both implicit and explicit feedback for a single feature.

6. Align feedback to improve the AI. Your feedback mechanisms should help improve and tune your AI model.

7. The process of designing feedback mechanisms should be collaborative across product functions. Extend your design team to include product managers, engineering and machine learning teams, and UX counterparts.

8. The process of collecting feedback is the process of enabling users to teach the AI. Feedback may be fed directly into the AI model, or your team can collate feedback, analyze it, and make changes later.

9. When collecting feedback, ensure that feedback mechanisms don't get in the way of core user workflows or stop users from doing their job.

10. The value of giving feedback is tied to motivation. Users are more likely to give feedback if it gives them direct or indirect benefits. Consider user motivations like reward, utility, altruism, and self-expression when designing feedback mechanisms.

11. When users provide feedback to your AI, they need
     to know that the system received the feedback.
     Your interface should confidently respond to user
     feedback.

12. Give people control over the feedback they give
     your AI. If people feel more in control, they are more
     likely to trust your system. Allow users to see and
     edit feedback already given by them.

13. Instead of thinking of an AI system as something
     that operates in a silo, people and AI can be
     symbiotic partners in a system. AI systems
     are continuously learning. AI can give people
     superpowers, while people can teach and improve it
     through feedback.

CHAPTER 6

Handling Errors

This chapter will look at different types of errors that occur in AI systems and strategies for handling them.

When I was in university, I wanted to get a new laptop. My mom agreed to get it for my birthday and gave me a budget of about a thousand dollars. I started researching the best laptop I could get within the budget. I compared new and used laptops. The used ones were much cheaper, and I found one person on an online classifieds website selling the model I wanted for about four hundred dollars. This was less than half my budget. So I thought it would be worth checking out.

I reached out to the person, and they responded enthusiastically that the laptop was available. But I had to pay two hundred dollars as an advance payment, and I transferred that amount from my meager college savings.

After that, there was no response. A few days went by, and I reached out to them again. They said there was some issue with the delivery provider, and they could expedite it if I could transfer them a hundred dollars more. This was already a red flag, but I was so impatient and jumpy to get the laptop that I sent the money. Once again, there was no response. I reached out, and they said that there was some other problem and they wanted me to transfer more money. I told them I would not give them any more until I saw the laptop in person. They stopped replying altogether and even deleted their classifieds account. I lost all the money I had saved that year in just a few days.

© Akshay Kore 2022
A. Kore, *Designing Human-Centric AI Experiences*,
https://doi.org/10.1007/978-1-4842-8088-1_6

I got scammed, and it was too embarrassing to tell anyone.

After a few weeks, I gathered the courage to tell my mom about this incident, and she was understandably furious. We went to the police station to file a complaint, but nothing really came of it, and I never got my money back.

Now I try my best not to get fooled. If a deal seems too good to be true, it probably is. Because of this incident, I never make large purchases from unknown sellers or websites. I lean toward paying in person for the most significant purchases, even if it might cost more. I prefer the peace of mind over the anxiety of getting scammed again. As promised, my mom got me the laptop I wanted for my birthday, which was bittersweet.

Errors happen. Mistakes happen. It is important to recover effectively and learn from them when they do. When a person makes a mistake, you lose a little trust in them. But they can regain the lost trust by acknowledging the error, apologizing for it, making amends, and learning from the mistake. In school, children learn by trying to solve a problem, making mistakes, and getting feedback. Correction of errors can make things better. At my workplace, we frequently conduct design reviews to point out errors in user experience workflows, give feedback to each other, and improve the product. Mistakes are inevitable when we try to learn, build, or improve something.

Errors Are Inevitable in AI

AI systems are not perfect; they are probabilistic and will be wrong at some point. Errors in your AI system are inevitable, and you need ways to handle them. Your users will test your product in ways you can't foresee during the development process.[1] Misunderstandings, ambiguity, frustration, and mistakes will happen. Designing for these cases is a core user experience

---

[1] pair.withgoogle.com, https://pair.withgoogle.com/.

problem. Errors can also lead to prospects for improving your product through feedback. Errors can also be opportunities for explaining how the AI system works. They can help establish the correct mental models and recalibrate user trust.

Errors can be opportunities for building trust.

Humble Machines

Learning, in AI or otherwise, can't happen without making mistakes. People are more likely to forgive an AI that acknowledges that it is learning and will sometimes be wrong. An AI system that assumes it knows everything will be less trustworthy when it makes mistakes.

Design your AI products with the knowledge that errors are an integral part of the user experience—acknowledging that your AI product is learning and continuously improving will help you create opportunities for dialog with users through interactions, explanations, or feedback. Humility is a feature.

Design your AI products with the knowledge that errors are an integral part of the user experience.

Guidelines for Handling AI Errors

When building an AI product, error handling is a critical design problem. A system that doesn't manage its mistakes well can be frustrating to use and erode the user's trust in the AI. Designing effective error states and messages can help you regain user trust and establish the correct mental models.

The following are some considerations when handling AI errors:

1. Define "errors" and "failures."

2. Use feedback to find new errors.

3. Consider the type of error.

4. Understand the stakes of the situation.

5. Indicate that an error occurred.

6. Don't blame the user.

7. Optimize for understanding.

8. Graceful failure and handoff.

Define "Errors" and "Failures"

Understand what type of mistake is an error and what constitutes a failure. What the user considers an error is deeply connected to their expectations of the AI system.[2] A music streaming service that predicts irrelevant answers 20% of the time can be viewed as an error, while an autonomous car that crashes 20% of the time, that is, one out of five times, is dangerous—it is a system failure. The line between error and failure is fuzzy. Failure for a music app might be no response from the AI; an error might be an incorrect prediction. Whether a mistake is an error or failure will depend on the context of the environment and the stakes of the situation. You should tolerate errors, but not failures. Failures make the system unusable; an error is an inconvenience.

Use Feedback to Find New Errors

While you might be able to predict some errors during the product development phase, it is best to use quality assurance exercises and pilots to find potential errors initially. After your AI is deployed, continue monitoring feedback to discover new errors throughout the product life cycle and as you launch new features. Collaborate with your team

---

[2] pair.withgoogle.com, https://pair.withgoogle.com/.

to decide on some channels to discover additional errors from users. Consider the following sources:[3]

1. Customer service reports

2. Customer service tickets

3. Comments on social media

4. App store reviews

5. In-product metrics

6. In-product surveys and explicit feedback mechanisms

7. Observing user behavior

8. User interviews

As you discover new errors, work with your team to design the correct error handling mechanisms and identify problems that need to be solved.

Consider the Type of Error

While your AI product will encounter common errors that occur in non-AI systems, some types of errors are unique to probabilistic systems like AI. What constitutes an error can differ from user to user based on their expertise, goals, mental models, and past experience using the product.[4] The type of error can also depend on its source.

The following are some common types of errors you will encounter in AI systems:

1. System errors

2. User errors

3. User-perceived errors

---

[3] pair.withgoogle.com, `https://pair.withgoogle.com/`.

[4] pair.withgoogle.com, `https://pair.withgoogle.com/`.

System Errors

These are errors that occur due to inherent limitations of the AI system like lack of adequate training data, bias in existing data, incorrect models, bad network, corrupt data, etc. When system errors are encountered, your users will likely blame your product for not meeting their needs.

User Errors

These errors happen when users make a mistake when interacting with the AI, like giving incorrect inputs, pressing the wrong buttons, malicious intent, etc. When user errors occur, your internal teams will likely blame the user for making a mistake. Be careful not to blame users when such errors happen.

User-Perceived Errors

User-perceived errors are unique to probabilistic systems like AI. This is when something feels like an error to the user, but it is not an error for the AI. It occurs due to a mismatch between the AI's results and the user's expectations. Also known as context errors, they can emerge if your product assigns incorrect weights to different preferences. For example, a user liking a video of a butter chicken recipe might expect to see more Indian recipes, but your model recommends more videos by the same creator. Users might get confused or stop using the product since it doesn't meet their expectations. To handle context errors, you'll need to evaluate if the AI makes the correct assumptions by assigning appropriate weights to different inputs. User-perceived errors can also happen when the system presents low-confidence results or fails to provide a result altogether.

(a) Netflix match percentage

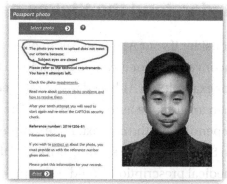

(b) Robot passport checker rejects Asian man's
application because "eyes are closed."

Figure 6-1. User-perceived errors. *(a) Netflix suggestions sometimes
don't match user expectations. In this case, the user expects this title
to have a higher match percentage than indicated. Source: Netflix
website on desktop. (b) A passport-checking software rejects an Asian
man's application because it thinks their eyes are closed. Source:
www.telegraph.co.uk/technology/2016/12/07/robot-passport-
checker-rejects-asian-mans-photo-having-eyes/*

Table 6-1. *Types of errors*

Type of error	Error subtype
System errors	Data errors
	Relevance errors
	Model errors
	Invisible errors
User errors	Unexpected or incorrect input
	Breaking user habits
User-perceived errors	Context errors
	Failstates

Understand the Stakes of the Situation

Consider if your AI mistake is an error or a failure. That is, is it an inconvenience or a system breakdown that can have severe consequences? For example, an AI-based restaurant recommender has different stakes than an airplane autopilot system. The risk of an error can change depending on the context. For example, a quick message to your colleague on Slack with some spelling errors is OK, but a similar spelling error in a medical prescription would be hazardous. A user might say "No symptoms of hematoma," while the system outputs "Symptoms of hematoma" because it missed a word, which completely changed the output. In case of a medical prescription error, you might want the user to double-check the spelling.

When designing your error responses, consider the stakes of the situation. Scenarios like health, safety, finance, or sensitive social contexts are high stakes, while play, experimentation, entertainment, or nonessential recommendations are low stakes.

Indicate That an Error Occurred

Explicitly indicate that something went wrong. The very worst error messages are those that don't exist. When users make mistakes and get no feedback, they're completely lost.[5] A case where a user notices an error but the system gives no indication can erode user trust when using your AI product in the future.

---

[5] Nielsen, Jakob. "Error Message Guidelines." Nielsen Norman Group, 2001, www.nngroup.com/articles/error-message-guidelines/.

Don't Blame the User

Your error messages should be polite, precise, and constructive. Don't blame the user even if the error occurred because of the user's mistake. Don't imply that the user is stupid or doing something wrong, like "Incorrect input" or "Illegal command." Phrase messages politely like "We can't understand this right now. Can you check the entered information and try again?" Instead of accusing the user of an error, indicate that an error occurred and provide helpful advice on what caused it and how to correct it.

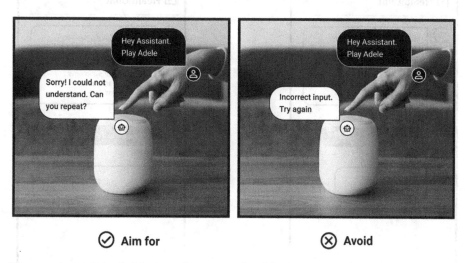

Figure 6-2. Don't blame the user. (Left) Example of a polite, precise, and constructive response. (Right) Giving a robotic response can sound rude to some users

Optimize for Understanding

Your AI error messages should be understandable by your stakeholders. They should be brief and to the point. Avoid explaining anything that is not related to the current context. Use human-readable language instead of obscure codes that only your team understands, like

"Exception: Error 420 occurred." Precisely describe the exact problem instead of vague explanations like "System error." Reduce ambiguity, and make error messages understandable and clearly visible. For example, when your system can't connect to the Internet, it is better to say "Cannot connect to the Internet right now. Please refresh or try again later" rather than saying "Error 404 occurred."

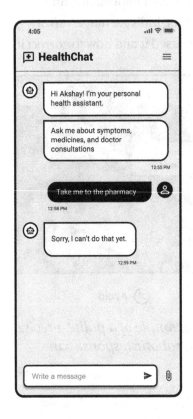
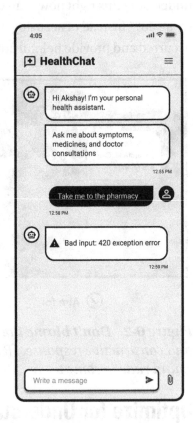

 **Aim for**    **Avoid**

Figure 6-3.   Optimize for understanding. (Left) Make error messages understandable and clearly visible. (Right) Avoid obscure codes

Graceful Failure and Handoff

Errors will happen. The trick isn't to avoid errors, but to find them and make them just as user-centered as the rest of your product.[6] Allow users to move forward when your AI makes a mistake. Give users a path forward depending on the severity of the error. Address the error at the moment: explain what happened, why it happened, what they can do about it, and how they can prevent it from happening in the future.

The following are some considerations for graceful failure and handoff:

1. Provide appropriate responses.

2. Use errors as opportunities for explanation.

3. Use errors as opportunities for feedback.

4. Disambiguate when uncertain.

5. Return control to the user.

6. Assume intentional abuse.

Provide Appropriate Responses

When something goes wrong, address the error at the moment. You can also let users know that there is a recovery plan in case of an error. Advanced knowledge can help users set the right expectations and make more informed decisions. For example, a music streaming service can become unusable when there is no Internet. Communicating that the service will automatically play downloaded songs when there is no network connection can help users feel more comfortable and informed about the system's behavior.

---

[6] pair.withgoogle.com, `https://pair.withgoogle.com/`.

Use Errors as Opportunities for Explanation

When an error occurs, it can be an opportunity to explain how the system works, its limits, and its capabilities. It can help you set the right user expectations for your AI. Enable users to understand why the AI behaved as it did. Sometimes the system is working as intended, but users perceive it to be an error because there is a mismatch between the user's understanding of how the system works and how it actually works. Error explanations can be a way of bridging this gap in understanding.

Your error messages can be an opportunity to educate users. Most users of your product won't read documentation, especially if it is not essential to their task, and they read documentation only when they are in trouble. Given this, you can use error messages as an educational resource to impart a small amount of knowledge to users.[7] Don't explain everything; your error messages should be brief and to the point. Sometimes you can use hyperlinks to connect short error messages to detailed explanations. But avoid overdoing this.

Your AI can teach users how they can leverage its capabilities better. Your error messages can teach users how your AI system works, establish the correct mental models, and give them the information they need to use it better.

---

[7] Nielsen, Jakob. "Error Message Guidelines." Nielsen Norman Group, 2001, www.nngroup.com/articles/error-message-guidelines/.

Figure 6-4.   Use errors as opportunities for an explanation. You can use errors to explain how the system works, its limits, and its capabilities, and they can help you set the right user expectations of your AI. In this example, the error message indicates that the system works on plants native to India and the conditions in which the system works best

Use Errors as Opportunities for Feedback

When your AI system makes mistakes, you can learn from them by asking for feedback. Include opportunities for users to provide feedback both when presented with an error message and alongside "correct" system output,[8] for example, allowing users to like or dislike AI suggestions in case of user-perceived errors.

---

[8] pair.withgoogle.com, https://pair.withgoogle.com/.

Think of errors as opportunities for feedback. Giving feedback lets users feel more in control of the product when errors happen, thereby calibrating the right trust level. When an AI acknowledges its mistakes and asks users to help it improve, users are more likely to provide feedback. Asking for feedback also reinforces the learning nature of your AI product and helps users adjust their mental models.

⟨⟩ Dog detector

Figure 6-5. Use errors as opportunities for feedback. The dog detector application misclassifies a muffin as a chihuahua in this example. Along with the prediction, the application allows the user to provide feedback. Image source: www.huffpost.com/entry/ blueberry-muffin-chihuahua_n_4649064

Disambiguate when Uncertain

Disambiguation means removing uncertainty from meaning, that is, ambiguity. In the context of your AI errors, it can mean providing multiple suggestions to reduce the perception of AI error. For example, when someone is looking for a Greek eatery in a restaurant recommendation service and the AI can't find any with high confidence, it might recommend restaurants with similar cuisines like Lebanese or Turkish. Or when you tell a voice assistant to play "God Is a DJ," it asks whether it should play "God Is a DJ" by Faithless or Pink.

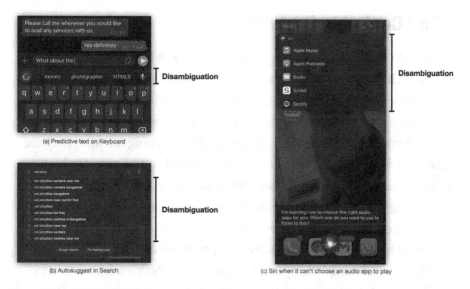

Figure 6-6. Disambiguate when uncertain. (a) Predictive text on Google Keyboard shows multiple suggestions for the user to choose from. Source: Gboard on iOS. (b) Google Search's autosuggest presents users with multiple options. Source: Google Search. (c) Apple's Siri suggests multiple applications to play an audio request when uncertain. Source: Siri on iOS

Gracefully degrade by disambiguating your AI's suggestions when uncertain. This is especially useful in low-confidence cases where users perceive the AI as less accurate.

Whether you should disambiguate or not will depend on the stakes of the situation. It is OK to disambiguate for low-stakes scenarios like movie or restaurant recommendations. But for high-stakes situations like self-driving vehicles, disambiguation can become confusing or even dangerous. You won't want the car to show you options of possible actions if there is a child in front of the vehicle, and you would expect the car to decide to stop immediately.

Return Control to the User

When an AI makes an error, the easiest option is to hand off control to the user. When letting users take over the system, consider if providing a full override would be appropriate. However, this depends on the stakes of the situation. In low-stakes cases like recommending funny videos, users can take over immediately by manually searching for what they are looking for. However, reverting to manual control can be risky for high-stakes scenarios. For example, an autonomous car suddenly asking for the passenger to take over when it encounters an unknown object like a diversion on the road would be extremely dangerous. When this AI-to-manual handoff happens, it's your responsibility to make it easy and intuitive for users to quickly pick up where the system leaves off.[9] Users must have all the information they need to take over: awareness of the situation, immediate next steps, and instructions on how to do them.

---

[9]pair.withgoogle.com, `https://pair.withgoogle.com/`.

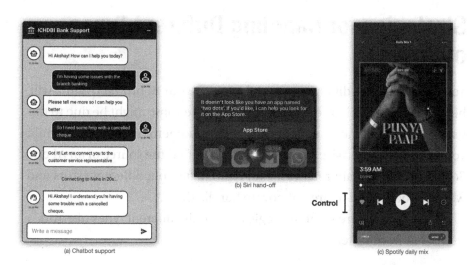

(a) Chatbot support   (b) Siri hand-off   (c) Spotify daily mix

Figure 6-7.   Return control to the user. *(a) Example of a banking chatbot that gracefully hands off to a human agent when it can't answer a specific question. (b) Apple's Siri asks the user to choose an application from the app store when it can't complete a request. Source: Siri on iOS. (c) Users can easily skip, go back, or forward songs on personalized daily playlists on Spotify. Source: Spotify app*

Assume Intentional Abuse

While it is important to design understandable error messages with appropriate responses, you should assume that certain users will intentionally abuse the system. Assume subversive use and make error responses safe, boring, and a natural part of the user experience. Avoid making dangerous failures interesting or overexplaining system vulnerabilities—that can incentivize some users to reproduce them.[10] For example, an AI-based cybersecurity system that defends against hackers can be broken into if it overexplains exactly how it prevents attacks.

---

[10] pair.withgoogle.com, https://pair.withgoogle.com/.

Strategies for Handling Different Types of Errors

You will encounter different types of errors in your AI product. Some will be caused by faults in the system or data, while some might be due to the user's mistake. Sometimes users might think something is an error, while your AI thinks otherwise. It can be difficult to find the source of an AI error. So you need to collaborate with your team to find out different errors and why they occur and figure out how to handle them.

The following are some strategies to handle different types of errors that occur in AI systems.

System Errors

These errors occur due to a fault in how the AI system is designed. System errors can happen due to an incorrect or inaccurate AI model or deficiencies in data. This can be due to an inherent system limitation when the AI can't provide a good answer or any answer. While system errors can also occur in non-AI products, some system errors are unique to AI products. A general strategy to deal with a system error is acknowledging the mistake and asking for feedback.

The following are some types of system errors:

1. Data errors

2. Relevance errors

3. Model errors

4. Invisible errors

Data Errors

These types of system errors occur due to limitations in training data or how the AI model is tuned. These result in incorrect or irrelevant predictions. For example, a recipe suggestion app trained on Indian cuisine will not be able to help users when they ask questions about Italian food. The following are some sources of data errors.

Mislabeled or Misclassified Data

These errors occur when the training data is labeled or classified incorrectly, resulting in poor results. For example, in an app that recognizes different dog breeds, mislabeling a golden retriever as a labrador can lead to misclassification.

Error Resolution

The best way to handle errors due to mislabeled data is to ask users to correct the system through feedback mechanisms. You can ask users to provide the correct suggestion or simply indicate that the suggestion is wrong. We can then give this feedback to the product development team to analyze and update the model, or it can be directly fed into the AI to improve the dataset.

📷 Dog detector

Figure 6-8. Mislabeled or misclassified data. *The dog detector application misclassifies a pug as a chihuahua in this example. Along with the prediction, the application allows the user to provide feedback. Image source: Photo by Ashleigh Robertson on Unsplash*

Incomplete Data

These errors occur when some parts of the training data are incomplete, leading to a mismatch in user expectations. For example, a healthcare AI assistant might answer questions about heart ailments but not kidney stones since it was not trained on that data.

Error Resolution

Clearly explain what your AI can do and what it can't. Indicate types of inputs and environments it works best in, its limitations, and what's missing. Additionally, allow users to give feedback about their expectations of the system.

Figure 6-9. Incomplete data. In this example, the error message indicates the system's limits by mentioning that it works on plants native to India and the conditions in which the system works best. Additionally, the interface also allows users to give feedback

Missing Data

An AI system trained on one type of data might not work effectively for a different type of data. System errors due to missing data happen when the user reaches the limits of what your AI is trained to do. For example, if a user tries to use a dog classification app on a person, the system won't provide the right results.

Error Resolution

Clearly explain what your AI can do and what it can't. Indicate types of inputs and environments it works best in, its limitations, and what's missing. Additionally, allow users to give feedback about their expectations of the system.

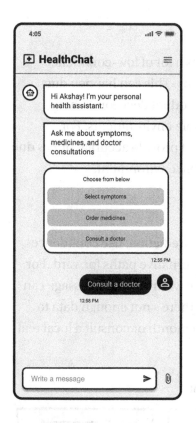

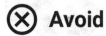

⊘ **Aim for**   ⊗ **Avoid**

Figure 6-10. Missing data. (Left) Aim to explain what the AI can do. In this example, the bot indicates its capabilities and boundaries. (Right) Avoid open-ended statements. In this example, saying "Ask me anything" is misleading since users can't ask anything they want

Relevance Errors

Your AI will not be able to provide useful predictions all the time. Relevance errors occur due to a fault in the design of the AI model or problems with training data. These errors result in the AI's output appearing irrelevant to the user's current needs. The following are some sources of relevance errors.

303

Low-Confidence Results

This error occurs when the AI system suggests a lot of low-confidence results and feels too uncertain of its predictions. This can happen due to a lack of data, incomplete data, incorrect product decisions when determining confidence thresholds, or unstable environments. For example, a home price prediction service can't provide accurate prices due to an economic recession, which is an unstable environment.

Error Resolution

Clearly explain why the AI made a particular prediction, use confidence values to inform UI decisions, and provide alternative paths forward. For example, in the home price prediction service, your AI error message can say, "Due to changing economic conditions, there's not enough data to predict home prices right now. Try again in a month or consult a local real estate advisor."

🎤 **Voice Assistant**

High confidence   **Medium confidence**   **Low confidence**

Figure 6-11. *Low-confidence results. Based on the confidence level, the voice assistant provides different responses*

Irrelevance

Sometimes your AI predictions can be completely irrelevant to the user even if your AI predicts them with high confidence. This can lead to a mismatch in user expectations and erode trust. Irrelevance can happen due to faults in the AI algorithm or problems with the training data. For example, a ride-hailing service recommends vacation stays when the user has booked a taxi to the hospital.

Error Resolution

When the system presents irrelevant results, allow users to provide explicit feedback. This feedback can then be given to the product team to analyze and update the model, or it can be directly fed into the AI to improve the dataset. You can also make it easy to ignore the suggestion.

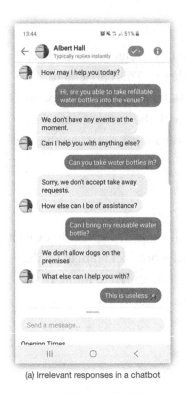

(a) Irrelevant responses in a chatbot

(b) Google lens thinks the person is a clock

Figure 6-12. Examples of irrelevance. *(a) Irrelevant responses from a chatbot can become annoying. Source:* www.userlike.com/ en/blog/chatbot-fails. *(b) Google Lens thinks that the person in the image is a clock. Source:* www.reddit.com/r/funny/comments/ au3l6y/google_lens_thinks_my_girlfriend_is_a_clock/

Model Errors

These errors happen due to faults in how the AI model is designed. Model errors can result in incorrect or irrelevant suggestions even if the underlying training data is accurate. The following are some sources of model errors.

Incorrect Model

These errors happen due to faults in how the AI model is designed. Sometimes despite adequate data, the AI model isn't precise enough. An incorrect model can lead to irrelevant, low-confidence results that are not useful. For example, an Italian food detection service has good data on various types of pasta but poor model training due to lack of rigorous testing or low-accuracy results in a large number of misclassifications between macaroni and penne.

Error Resolution

Ask users to correct the system through feedback mechanisms. You can ask users to provide the correct suggestion or simply indicate that the suggestion is wrong. This feedback can then be given to the product team to analyze and update the model.

(a) Netflix thumbs up/down   (b) Grammarly word suggestion

Figure 6-13. Incorrect model. (a) Like and dislike actions on Netflix titles enable users to give feedback on incorrect suggestions. Source: Netflix website on desktop. (b) Grammarly allows users to flag incorrect suggestions. Source: Grammarly plugin on desktop

307

Miscalibrated Input

There are times when your AI model has sufficiently precise data and works reasonably well on your expected use cases, but your system improperly weighs user inputs, actions, and preferences. Miscalibrated inputs can lead to a mismatch between AI outputs and user expectations. For example, after watching a butter chicken recipe, a video streaming service recommends more videos of butter chicken, but the user expects more content from the same creator.

Error Resolution

Explain how the system matches inputs to outputs and allow the user to correct the system through feedback.[11]

Netflix suggestions

Figure 6-14. Miscalibrated input. Like and dislike actions on Netflix titles enable users to give feedback. The system also explains how the suggestions were derived. Source: Netflix website on desktop

Security Flaws

While using your product, some users might discover security flaws or ways to hack your AI model. This discovery can be accidental; it can be malicious actors trying to exploit your system or researchers trying to

---

[11] pair.withgoogle.com, https://pair.withgoogle.com/.

expose security flaws to help your team fix them. For example, a group of researchers developed small, inconspicuous stickers that could be placed on a traffic sign, resulting in a computer vision system—similar to those used in self-driving cars—to misclassify the sign.[12]

Error Resolution

While these errors can't be detected easily, encourage users to provide feedback to your team, especially if your AI operates in high-stakes situations like self-driving cars or healthcare. You can also organize a bug bounty program to encourage ethical hackers to find and report vulnerabilities and security flaws in your AI.

(a) Stickers placed on stop signs

(b) Spectacles that fool AI

Figure 6-15. Hacking AI models. (a) Specialized stickers created to fool AI systems into predicting "stop sign" as "speed limit 40 m/hr." This could fool an autonomous vehicle reliant on image data. Source: www.oreilly.com/library/view/strengthening-deep-neural/9781492044949/ch01.html. (b) Researchers created spectacles that fool face recognition systems into predicting an incorrect output. In this example, the man wearing these spectacles is predicted as the actress Milla Jovovich. Source: www.dailymail.co.uk/sciencetech/article-3898508/Spot-difference-facial-recognition-systems-t-bizarre-face-stealing-specs-fool-AIs-thinking-else.html

---

[12] Mitchell, Melanie. *Artificial Intelligence*. First ed., Farrar, Straus and Giroux, 2019.

Invisible Errors

These are errors that users don't perceive. Since these errors are invisible to the user, you don't need to worry about how to explain them in the user interface, but being aware of them could help you improve your AI.[13] If undetected for a long time, these errors can have significant consequences for the user and have important implications for how your system measures error rates and failures. The following are a few types of invisible errors.

Background Errors

They happen in the background where your team recognizes them, but neither the user nor the system registers an error. For example, a search engine provides an incorrect result with high confidence, but the user doesn't find a mistake.

Error Resolution

Since you are unlikely to detect background errors through system error logs or user feedback, you'll need to find these errors by stress-testing the system. You can incorporate a quality assurance process and give feedback to your product team to fix the issue. When handling background errors, consider the stakes of the situation and the impact of the error on the user.

Happy Accidents

In these cases, the system perceives something to be a poor prediction, limitation, or error, but users find it interesting. For example, asking a smart speaker to cook dinner gives a funny response.

---

[13] pair.withgoogle.com, https://pair.withgoogle.com/.

Error Resolution

Error responses to happy accidents can be opportunities to delight users. You can use humor or imbue personality into your product experience. Be on the lookout for happy accidents by periodically observing users and testing the AI internally.

(a) Siri response

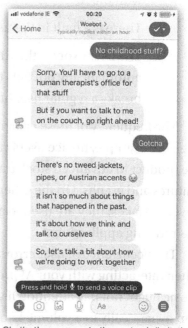

(b) Chatbot's response to the system's limitation

Figure 6-16.   Happy accidents. *(a) Funny response by Siri to the question "What is zero divided by zero?" Source:* `https://me.me/i/9-38-hey-siri-what-is-0-0-tap-to-edit-7ae45ca00977442` `2a392c55330908344.` *(b) A clever response by a chatbot to a system limitation. Source:* `https://uxplanet.org/having-a-laugh-humor-and-chatbot-user-experience-e6b7636a454d`

User Errors

User errors happen when there is a mismatch between the user's expectation of what should happen and what actually occurs. User errors can occur due to various reasons:

1. The user makes a mistake while interacting with the AI.

2. There is a fault in the user interface of your product.

3. The user expects the AI to correct their input automatically but fails to do so or when your UI breaks a regular user habit.

The frequency of these errors can depend on the expertise of the user. For example, novice users can make several mistakes when using the product in advanced stock trading applications with many complex features and graphs. On the other hand, advanced stock traders will make fewer mistakes; they are used to and expect complex features and charts in their trading applications.

The occurrence of user errors can also depend on users' attentiveness when interacting with your AI product. For example, a person listening to music might not pay attention if your AI changes the next button on the app dynamically to a volume knob. When they want to change the song, they might tap on the exact location, sometimes even without looking, but the volume changes instead, leading to a mistake. The following are some common sources of user errors.

Unexpected or Incorrect Input

Sometimes when users enter an incorrect input into the system, they expect the AI to understand the mistake and automatically correct it. This type of error happens if the AI doesn't autocorrect while the user anticipates a correction. For example, a user makes a typo in a search query and expects the system to recognize the intended spelling.

Error Resolution

Collect implicit feedback by observing user behavior. For example, a person manually corrects a typo or repeats a voice query multiple times until they are satisfied. Offer assistance by providing a range of possible answers (N-best lists) for incorrect, ambiguous, or unexpected input. For example, when a user enters the word "Italian" as a query, the search result can say, "Showing results for Italian (cuisine). Did you mean Italian (language) instead?"

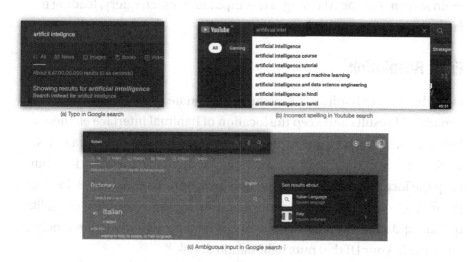

Figure 6-17.   Unexpected or incorrect input. (a) Google Search autocorrects typos and presents search results by the correct spelling. Source: Google Search on desktop. (b) YouTube offers autosuggest options even when the user makes a spelling mistake. Source: YouTube website on desktop. (c) When a user makes an ambiguous query like "Italian" that can have different intents like country, cuisine, or definition, Google Search helps users search by their preferred meaning like "Italian language" or "the country Italy." Source: Google Search on desktop

Breaking User Habits

For products used frequently, users start forming habits when interacting with them. You expect the position of a URL input to be constant on a browser, and you expect your car's steering wheel to be in the same place every time you enter. Predictable interfaces are necessary for habit-forming products or when the stakes are high. User errors can happen if your AI system interferes with users' habits when interacting with the product. For example, clicking your phone's home button takes you to the main screen, but the AI changes this input to a search query, leading to false starts, errors, and frustration.

Error Resolution

Don't break user habits unnecessarily. Designate specific areas for dynamic AI results and keep the location of habitual interface elements like search, settings, close button, etc. constant. For example, in a news application, you can dynamically change elements on the news feed but keep the location of the search bar and back buttons constant. If the user is coming to your product to perform important, time-sensitive tasks, like quickly updating a spreadsheet before a client presentation, don't include anything in your UI that puts habituation at risk.[14]

If changing user habits is necessary, allow users to give feedback or revert to, choose, or retain the original interaction pattern. For example, when Twitter changed the order of tweets in the feed to a dynamic sequence, it provided users with a fallback to revert to the original chronological timeline.

---

[14] Fisher, Kristie, and Shannon May. "Predictably Smart—Library." Google Design, 2018, https://design.google/library/predictably-smart/.

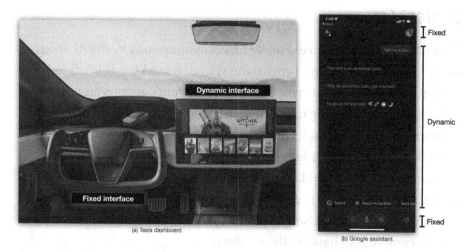

Figure 6-18. Don't break user habits. (a) Interiors of a Tesla differentiate between dynamic AI suggestions on the screen and dashboard and static controls like steering, acceleration, and brake pedals. Source: www.theverge.com/2021/1/27/22252832/tesla-model-s-redesign-plaid-mile-range-interior-refresh. (b) Habitual UI elements like microphone, text, profile, etc. are located in constant fixed places. Source: Google Assistant app

User-Perceived Errors

Sometimes an AI result can feel like an error to the user, but it isn't. User-perceived errors happen when there is a mismatch between user expectations and the system's assumptions. These types of errors are unique to probabilistic systems like AI. They occur when the system's actions and limitations aren't explained adequately or your product has not built the right mental models of your AI or because of poor assumptions by users. The following are some sources of user-perceived errors.

Context Errors

Context errors happen based on incorrect assumptions by the AI about user preferences. In context errors, the system is working as intended, but the user perceives an error because the system's actions aren't well-explained, break the user's mental model, or were based on poor assumptions.[15] For example, booking a flight on a travel website creates an event on the family calendar, which is unnecessary.

Context errors can make the AI system less useful by making incorrect assumptions about user preferences and needs, and they can erode trust in your AI significantly. Users might be confused or frustrated, fail at their task, or completely abandon the product.

Context can be related to personal preferences or broad cultural values. For example, a video streaming service that shows videos of a performer even when the user has repeatedly disliked similar videos can feel like an error. This happens when your product assigns incorrect weights to different user actions and preferences. For example, after liking a song by an artist, the user expects the system to suggest more songs from the artist, but instead, it starts recommending songs from the same genre.

Error Resolution

Ask users to provide explicit feedback on suggestions. This feedback can then be given to the product team to analyze and update the model or directly fed into the AI. Look at the weights your AI assigns to different signals, user actions, and preferences, and evaluate if they are incorrect, overvalued, underlooked, or missing. Addressing context errors will also depend on the frequency of use and stakes of the situation. Sometimes, you might decide to change the model to align with user needs and expectations. In other cases, it can be more helpful to establish better

---

[15] pair.withgoogle.com, https://pair.withgoogle.com/.

mental models by adjusting your onboarding and explanations. You can shift the user's perception of these situations from errors to expected behavior.[16] For example, a feedback mechanism can clearly ask users to show more songs by the same artist or genre.

(a) Netflix match percentage   (b) Grammarly word suggestion

Figure 6-19.   Context errors. (a) Netflix suggestions sometimes don't match user expectations. In this case, the user expects this title to have a higher match percentage than indicated. Like and dislike actions on Netflix titles enable users to give feedback. Source: Netflix website on desktop. (b) Grammarly allows users to flag suggestions. Source: Grammarly plugin on desktop

Failstates

Sometimes your AI product can't provide a satisfactory result or any result due to the system's inherent limitations. Failstates can happen if your system is not trained for what the user is asking it to do. For example, a voice assistant trained in the English language can't respond to commands in Hindi.

---

[16] pair.withgoogle.com, https://pair.withgoogle.com/.

Error Resolution

When the system fails to answer a question, incorporate appropriate empty states. Explain why it failed and describe the limitations of the system. Additionally, you can ask your users to provide feedback on the result, and this feedback can then be given to the product team to address the limitation.

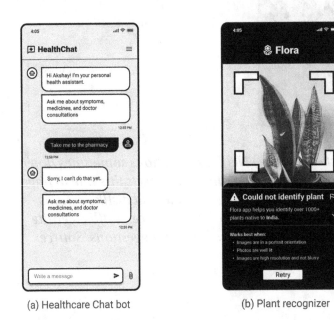

(a) Healthcare Chat bot   (b) Plant recognizer

Figure 6-20. Handling Failstates. (a) Example of a healthcare chatbot that clarifies its capabilities and boundaries when it fails to provide a result. (b) In this example, the error message indicates the application's limits by mentioning that the system works on plants native to India and the conditions in which the system works best. Additionally, the interface also allows users to give feedback

Table 6-2. *Types of errors and methods of resolution*

Type of error			Error resolution
System errors	Data errors	Mislabeled or misclassified data	Collect feedback.
		Incomplete data	Explain what your AI can and can't do; collect feedback.
		Missing data	Explain what your AI can and can't do, indicate limitations, and collect feedback.
	Relevance errors	Low-confidence results	Explain why the AI made the prediction.
		Irrelevance	Collect explicit feedback.
	Model errors	Incorrect model	Collect feedback.
		Miscalibrated input	Explain what your AI can and can't do; collect feedback.
		Security flaws	Collect feedback; encourage ethical hackers to find and report vulnerabilities and security flaws.
	Invisible errors	Background errors	Improve quality assurance in your product development process.
		Happy accidents	Opportunity to delight users; use humor.

(continued)

Table 6-2. (*continued*)

Type of error		Error resolution
User errors	Unexpected or incorrect input	*Collect implicit feedback.*
	Breaking user habits	*Don't break user habits unnecessarily. Designate specific areas for dynamic AI results and keep the location of habitual interface elements constant.*
User-perceived errors	Context errors	*Explain what your AI can and can't do; collect explicit feedback.*
	Failstates	*Explain what your AI can and can't do, indicate limitations, and collect feedback.*

Recalibrating Trust

When your AI system makes mistakes, users can be concerned about their ability to continue safely after the error has occurred. The nature and severity of the error and your product's ability to recover will impact user trust. Your users might trust your system less if it makes an error and does nothing about it. To recalibrate trust, address the AI's mistakes with humility. Empathize with users, communicate with the appropriate responses, and provide paths forward. You can use errors to explain how the system works and why it made a mistake and allow users to improve the AI through feedback. If your AI is prone to making mistakes in certain circumstances, you can hand over control to the user. Your error communication can be an opportunity to regain or prevent lost trust.

Summary

This chapter looked at different types of errors that occur in AI systems and strategies for handling them. Here are some notable points:

1. AI systems are not perfect; they are probabilistic and will be wrong at some point. Errors in your AI system are inevitable.

2. Errors can also be opportunities for improving your product through feedback and explaining how the AI system works. They can help establish the correct mental models and recalibrate user trust.

3. Design your AI products with the knowledge that errors are an integral part of the user experience.

4. Understand what type of mistake is an error and what constitutes a failure. You should tolerate errors, but not failures. Failures make the system unusable; an error is an inconvenience.

5. After your AI is deployed, continue monitoring feedback to discover new errors throughout the product life cycle and as you launch new features. Collaborate with your team to decide on some channels to watch to discover new errors from users.

6. Your AI error messages should be understandable by your stakeholders. They should be brief and to the point. Avoid explaining anything that is not related to the current context.

7. When an error occurs, it can be an opportunity to explain how the system works, its limits, and its capabilities.

8. When your AI system makes mistakes, you can learn from them by asking for feedback.

9. You will encounter different types of errors in your AI product. Some will be caused by faults in the system or data, while some might be due to the user's mistake. Sometimes users might think something is an error, while your AI thinks otherwise.

10. You need to collaborate with your team to find out different errors, why they occur, and how to handle them.

11. Consider the type of error. The following are some common types of errors you will encounter in AI systems:

    a. System errors

        These are errors that occur due to inherent limitations of the AI system. The following are some types of system errors:

        i.   Data errors

        ii. Relevance errors

        iii. Model errors

        iv. Invisible errors

    b. User errors

        These errors happen when users make a mistake when interacting with the AI. The following are some types of user errors:

      i.   Unexpected or incorrect input

      ii. Breaking user habits

  c. User-perceived errors

     User-perceived errors are unique to probabilistic systems like AI. This is when something feels like an error to the user, but it is not an error for the AI. The following are some types of user-perceived errors:

      i.   Context errors

      ii. Failstates

12. In general, you can handle errors by setting the right expectations, explaining what your AI can and can't do, collecting feedback, and providing clear, understandable responses. Refer to Table 6-2 for a summary of error handling strategies.

13. Your error communication can be an opportunity to regain or prevent lost trust. Empathize with users, communicate with the appropriate responses, and provide paths forward.

CHAPTER 7

AI Ethics

In this chapter, we will discuss the role of AI ethics in designing trustworthy, safe, and beneficial AI experiences.

In Mumbai, it is pretty common to travel for work or school by local train, and it is one of the fastest ways to get around the city. However, this speed comes with the tradeoff of large crowds and sometimes going through the entire journey without getting a place to sit. There are whole movies dedicated to this aspiration of getting a seat in a crowded Mumbai train. Frequent travelers often develop hacks for getting a seat by boarding from specific stations at particular times, making deals with seated passengers, or even risking their lives by jumping onto a moving train. Giving up your seat is a big deal when the stakes are so high.

Despite this, you often see people offering their seats to an elderly person. People have moral principles about what is right and wrong. Even though it is not mandatory to give up their seat, they offer it out of the goodness of their own heart. Morals are personal guiding principles by which people behave in a society. Morals are the basis for ethics. If you travel by the bus in Mumbai, you'll notice some seats reserved for the elderly and pregnant women. It is required for you to give up your seat to them. Ethics apply to society at large and are generally more practical. We codify and systematize moral behavior into ethical rules like reserving seats on public transport for pregnant women and the elderly or giving way to an ambulance.

> *AI ethics is a system of principles, values, and techniques intended to inform how we develop and use artificial intelligence responsibly.*

© Akshay Kore 2022
A. Kore, *Designing Human-Centric AI Experiences*,
https://doi.org/10.1007/978-1-4842-8088-1_7

AI is used in all kinds of important and varied applications, from recommending jobs, criminal sentencing, and airplane navigation systems to cybersecurity. Artificial intelligence will be used to make predictions and decisions that affect people and societies. Even if you don't design the system to work with people, it's got to eventually work with people, so you'd better think about people.[1] If we are going to let AI make decisions, it should be aligned with the ethics of the society in which it operates. AI ethics is a system of principles, values, and techniques intended to inform the development and responsible use of artificial intelligence. The ethical design of AI products is not just another form of technical problem-solving. You will need a diverse set of opinions about how your AI impacts users, stakeholders, and sometimes society at large. Everyone needs to sit down and have, as part of their design teams, people who are going to help them think more broadly about the unintended consequences of the systems they're building.[2] As designers and developers of AI systems, we hold a vast share of the collective influence, and it is imperative to understand the ethical considerations of our work.[3] AI ethics can also be about the principles of teams that design AI systems or the behavior of the AI itself.

Ethics-Based Design

Raising ethical concerns can feel like an impediment to a project. Most companies have a goal of improving profits, market share, or shareholder value. Sometimes this focus on growth can come at the expense of ethical consequences. Ethics can take a back seat, especially in fast-growing industries and companies. However, ignoring ethical implications can erode user trust in your AI. An AI that is not trustworthy is not useful.

---

[1] Ford, Martin R. *Architects of Intelligence*. Packt, 2018.
[2] Ford, Martin R. *Architects of Intelligence*. Packt, 2018.
[3] "IBM Design for AI." ibm.com, `www.ibm.com/design/ai/`.

Sometimes, you might be required by law to consider ethical implications. Encourage people to talk about ethical implications if they arise. Try to incentivize team members to raise ethical concerns. Those who advocate for ethical design within a company should be seen as innovators seeking the best outcomes for the company, end users, and society.[4] There are no prescribed models for incorporating ethical design in products. Leaders can facilitate that mindset by promoting an organizational structure that supports the integration of dialogue about ethics throughout product life cycles.[5] Existing product development processes can be good opportunities to get your team thinking about the ethical implications of your system. The transition points between discovery, prototyping, release, and revisions are natural contexts for conducting such reviews.[6] Your team members can highlight concerns, identify risks, raise "red" flags, or propose alternatives in these reviews.

An AI that is not trustworthy is not useful.

AI ethics is a large subject, and it would be difficult to discuss everything from the topic in this chapter. For this book, we will focus on a subset of ethical considerations relevant for designing AI products. These can be categorized into the following:

1. Ensuring your AI systems are trustworthy

2. Ensuring your AI is safe

3. Ensuring your AI systems are beneficial

---

[4] Ethically Aligned Design: A Vision for Prioritizing Human Well-Being with Autonomous and Intelligent Systems. First ed., IEEE, 2019.

[5] Ethically Aligned Design: A Vision for Prioritizing Human Well-Being with Autonomous and Intelligent Systems. First ed., IEEE, 2019.

[6] Ethically Aligned Design: A Vision for Prioritizing Human Well-Being with Autonomous and Intelligent Systems. First ed., IEEE, 2019.

Trustworthy AI

All successful relationships are built on trust. We don't make large purchases on websites we don't trust, we deposit our money with trusted institutions, and we tend to buy products from trustworthy brands. The AI products you build will make predictions and decisions for people. Users need to be able to trust your AI's decisions for them to use the product consistently. Imagine spending millions of dollars on a restaurant recommendation system only to find out that people don't trust your AI's suggestions and therefore don't use it.

Building trust goes beyond just the user interface. Your sales and marketing communications, R&D efforts, PR mentions, awards, fundraises, and customer testimonials all contribute to reinforcing trust in your AI. Trust is a critical consideration when building AI products. Ensure that your AI systems are explainable, fair, reliable, and inclusive and respect the privacy of their users and those they affect.

Explainable AI

We don't blindly trust people who can't explain their decisions in most cases. In school, for a mathematics exam, we were asked to show the steps followed to derive an answer. Showing how I derived an answer demonstrated that I understood the question and formed the right mental models. My teacher could use it to correct mistakes where appropriate and provide nuanced feedback instead of a binary response. You can often trust that people know what they are doing if they can explain how they arrived at an answer or a decision. This is not always the case for AI. Showing how AI systems arrive at a decision is not easy. Explainable AI is a field that focuses on getting AI systems to explain their decisions in a way that humans can understand. It is also known as "transparent AI" or "interpretable machine learning." While the field is gaining traction and progressing quickly, we have yet to build an AI system that can explain itself thoroughly in a human-understandable way.

Explainable AI is a field that focuses on getting AI systems to explain their decisions in a way that humans can understand.

Explainability and trust are interlinked. When AI systems are used to help make decisions that impact people's lives, it is particularly important that people understand how AI made those decisions. When people understand how an AI makes decisions, they can make more informed decisions by assigning the right level of trust in the system.

Black Box Models

A black box model is one where we only know the AI's inputs and outputs without knowing its internal workings. Most AI systems operate as black boxes. They use neural networks, massive collections of statistical weights, and activation functions. To the human eye, these are essentially jumbles of numbers that are constantly adjusted to account for new data. Even creators of such systems can only guess how they arrived at a particular decision. In these structures, knowledge and learning are represented in ways mostly indecipherable to human observers.[7] We don't always know what the AI learned. In certain high-stakes situations like criminal sentencing or detecting cancer, explaining why and how the AI made a decision is critical. In such cases, product creators and decision-makers should carefully consider the implications of using black box AI systems.

Transparency

People can't explain all their decisions and thought processes either. However, you have formed predictable mental models of how people behave and their decision-making processes over years of living. Similarly, while explaining how an AI arrived at certain decisions can be challenging, you can provide certain cues like information about how the system works and how it interacts with data.

---

[7] Husain, Amir. *The Sentient Machine*. Scribner, 2018.

Transparency in AI refers to the ability of an AI system to explain its inner workings, thereby setting the right expectations and forming the right mental models. While we have discussed strategies to incorporate transparency in the AI's user experience in Chapter 4, here are some general considerations:

1. Explain how your AI system works through in-product explanations and technical documentation.

2. Explain how well it works and in what situations. Also, explain when it doesn't work well.

3. Explain what data it was trained on and highlight any limitations in the training data. For example, a speech recognition system trained on an Australian English language dataset may not work satisfactorily for British speakers.

4. Ensure your explanations are easily understandable by your users and stakeholders. Simply sharing a list of billions of data points, lines of code, or complex calculations is not an explanation that is easy to grasp. Consider your user group and provide the right level of explanation.

We need to ensure that the AI's decisions are fair, do not put certain individuals at a disadvantage, or do not reinforce undesirable stereotypes.

Bias

A group of researchers trained an AI from text on the Internet to learn about words and their relationships.[8] You could ask the system to form an analogy by providing a sample. For example, if "Tokyo" is to "Japan,"

---

[8] Bolukbasi, Tolga, et al. "Man Is to Computer Programmer as Woman Is to Homemaker? Debiasing Word Embeddings." NIPS'16: Proceedings of the 30th International Conference on Neural Information Processing Systems, 2016, https://arxiv.org/pdf/1607.06520.pdf. Accessed November 25, 2021.

"New Delhi" is to what? The answer would be India. Another analogy could be a "man" is to "father" and a "woman" is to "mother," which is a reasonable answer. However, they also found that the AI learned unhealthy stereotypes. So when you ask it, "'Man' is to 'computer programmer' as 'woman' is to what?" the same AI would output the result as "homemaker." I think this result is disheartening. A less biased AI algorithm would understand that men and women can equally be computer programmers as they can be homemakers.

Gender stereotype *she-he* analogies.

sewing-carpentry	register-nurse-physician	housewife-shopkeeper
nurse-surgeon	interior designer-architect	softball-baseball
blond-burly	feminism-conservatism	cosmetics-pharmaceuticals
giggle-chuckle	vocalist-guitarist	petite-lanky
sassy-snappy	diva-superstar	charming-affable
volleyball-football	cupcakes-pizzas	hairdresser-barber

Gender appropriate *she-he* analogies.

queen-king	sister-brother	mother-father
waitress-waiter	ovarian cancer-prostate cancer	convent-monastery

Figure 7-1. Man is to computer programmer as woman is to homemaker? A group of researchers trained an AI to find analogies on a popular dataset only to find that it presented harmful gender stereotypes. Source: Bolukbasi, Tolga, et al. "Man Is to Computer Programmer as Woman Is to Homemaker? Debiasing Word Embeddings." NIPS'16: Proceedings of the 30th International Conference on Neural Information Processing Systems, 2016, `https://arxiv.org/pdf/1607.06520.pdf`

Algorithmic bias is the tendency of machine learning algorithms to produce inappropriately biased decisions about loans, housing, jobs, insurance, parole, sentencing, college admission, and so on.[9] Bias matters.

---

[9] Russell, Stuart. *Human Compatible.* Allen Lane, an imprint of Penguin Books, 2019.

As AI systems are used to make critical decisions that affect people and
societies, we need to ensure that these decisions are fair, do not put certain
individuals at a disadvantage, or do not reinforce undesirable stereotypes.
In most cases, bias in AI algorithms is not intentional, and it occurs
because of faults in the underlying data on which the AI was trained.

Biased data is often a reflection of the circumstances and culture
that produce it. The more data you have, the worse this problem can get.
Bigger datasets don't necessarily eliminate bias. Sometimes, they just
asymptotically zero in on the biases that have always been there.[10] Of
course, these biases in AI training data reflect biases in our society, but
the spread of real-world AI systems trained on biased data can magnify
these biases and do real damage.[11] Eliminating bias is a hard problem.
Sometimes biases can creep in even when you have introduced the right
checks in your algorithms. An AI-based resume scanning system that is
explicitly prevented from "knowing" about things like gender and race
can still be biased if it learns to penalize resumes that include the word
"women's," as in "women's chess club captain."[12] Bias can also emerge
in information processing. For instance, speech recognition systems are
notoriously less accurate for female speakers than for male speakers.[13] We
can also unintentionally introduce bias in the way in which we present
AI systems. For example, a vast majority of humanoid robots have white
"skin" color and use female voices.[14]

---

[10] Polson, Nicholas G., and James Scott. *AIQ*. Bantam Press, 2018.

[11] Mitchell, Melanie. *Artificial Intelligence*. First ed., Farrar, Straus and Giroux, 2019.

[12] Dastin, Jeffrey. "Amazon Scraps Secret AI Recruiting Tool That Showed Bias
Against Women." Reuters, 2018, `www.reuters.com/article/us-amazon-com-jobs-automation-insight-idUSKCN1MK08G`.

[13] Ethically Aligned Design: A Vision for Prioritizing Human Well-Being with
Autonomous and Intelligent Systems. First ed., IEEE, 2019.

[14] Ethically Aligned Design: A Vision for Prioritizing Human Well-Being with
Autonomous and Intelligent Systems. First ed., IEEE, 2019.

In most cases, bias in AI algorithms is not intentional. It occurs because of faults in the underlying data on which the AI was trained.

Percentage of women in top 100 Google image search results for CEO: 11%
Percentage of U.S. CEOs who are women: 27%

Figure 7-2. *Incident of a biased search result for the term "CEO" on Google search. Source: www.washington.edu/news/2015/04/09/ whos-a-ceo-google-image-results-can-shift-gender-biases/*

Current AI systems do not have a moral sense of what is right and wrong. We need to be careful about letting AI make critical decisions for people, especially if we know of any underlying bias. For example, criminal sentencing algorithms have been notorious for being biased even when they are explicitly prevented from knowing the defendant's race or gender.[15] I would prefer an imperfect human judge with morals over a machine that doesn't understand what it is doing.

Biased data is often a reflection of the circumstances and culture that produce it.

Facial Recognition

Facial recognition systems raise issues that are fundamental to human rights like privacy and freedom. Several critical applications, such as security and surveillance, airport screening, identifying people in credit

---

[15] Polson, Nicholas G., and James Scott. *AIQ*. Bantam Press, 2018.

card transactions, and accessing government services, employ facial recognition algorithms. While the loss of privacy is a major concern, a larger worry is the reliability of these systems. It can seem creepy if the software can recognize people in a large crowd. Bias in facial recognition systems can be especially dangerous. If your face is matched in error, you might be barred from a store or an airplane flight or wrongly accused of a crime.[16]

Many commercial face recognition systems tend to be more accurate on white males than females or non-white individuals. One researcher pointed out that a widely used dataset for training face recognition systems contains faces that are 77.5% male and 83.5% white.[17] For example, a passport application AI rejected an Asian man's photo because it insisted his eyes were closed.[18] Camera software is prone to missing faces with dark skin or classifying an Asian face as blinking.[19] Even small errors can have large societal repercussions.

Sometimes these biases are subtle and can be hard to find. A photo of a male nurse in a hospital might be wrongly classified as female if the dataset has a large number of women in the role of a nurse. Such biases are apparent after they occur but are difficult to predict ahead of time.

---

[16] Mitchell, Melanie. *Artificial Intelligence*. First ed., Farrar, Straus and Giroux, 2019.

[17] Mitchell, Melanie. *Artificial Intelligence*. First ed., Farrar, Straus and Giroux, 2019.

[18] Ethically Aligned Design: A Vision for Prioritizing Human Well-Being with Autonomous and Intelligent Systems. First ed., IEEE, 2019.

[19] Rose, Adam. "Are Face-Detection Cameras Racist?" time.com, 2010, http://content.time.com/time/business/article/0,8599,1954643,00.html.

Figure 7-3. *Example of a camera face detection program identifying an Asian face as "blinking." Source:* https://thesocietypages.org/socimages/2009/05/29/nikon-camera-says-asians-are-always-blinking/

Causes of Bias

Bias in AI systems can be harmful and can have real-world implications on people's lives. The following are two of the most common reasons for AI bias:

1. Bias in training data

2. Lack of team representation

Bias in Training Data

The majority of the bias in AI systems is a reflection of the underlying training data. This bias in the AI training data often reflects preexisting biases in society. In most cases, people designing these systems don't intend them to be biased. Unfairness can creep into systems when the data isn't representative of the population. Take the example of AI-based criminal sentencing. If a particular neighborhood is policed a lot more than another (deliberately or not), you will have more sample data for one over the other. This would have an impact on predictions about crime. The actual collection itself may not have shown any bias, but because of oversampling in one neighborhood and undersampling in another, the use of that data could lead to biased predictions.[20] Another example could be lending, where, historically, a certain section of society had access to credit much before another, leading to a bias in the underlying training data.

We don't have representative amounts of data for people from different cultures, languages, and societal norms. Bigger datasets don't necessarily eliminate bias. Sometimes, more data can lead to more bias.

Eliminating AI bias is a difficult problem. Sometimes even ensuring "representativeness" isn't enough. The representative data can itself be racist and sexist. Training an AI system on such data may inadvertently lead to results that perpetuate these harmful biases.[21]

Lack of Team Representation

Bias can occur even if the team designing the system isn't necessarily biased. It is highly unlikely that a team of individuals belonging to the same race, same socioeconomic background, and similar life experiences

---

[20] Ford, Martin R. *Architects of Intelligence*. Packt, 2018.

[21] Smith, Brad, and Harry Shum. *The Future Computed*. Microsoft Corporation, 2018.

can design an unbiased system for a different persona without ever interacting with the user group. Inevitably unconscious biases will creep in.

Reducing Bias

Eliminating bias is a complex and hard task. It is tough to get rid of bias completely, and some level of bias will inevitably creep in. But we can always reduce and manage bias in our AI's datasets and outputs. The following are some suggestions for reducing bias:

1. Ensure that your AI's training data is balanced in its representation of its users. While your raw training data can itself be biased, make sure to calibrate it for appropriate representation of user demographics like gender, race, ethnicity, socioeconomic status, etc.

2. Your team should audit your AI's datasets and results regularly. Bias can creep in over time as your AI learns from its environment. Schedule team reviews to monitor results. Making sure that you check your AI's data and results regularly can help you identify and fix new cases of bias sooner.

3. Collect feedback about your AI's results actively from users and other stakeholders. Identify when bias has occurred and correct it.

4. Include diverse stakeholders in your design process. Make sure you test your product with relevant stakeholders throughout the product development process. Stakeholders can be users of your AI system, decision-makers, affected users, as well as

regulators. Include adjacent teams like customer
success, marketing, and experts like social scientists,
ethicists, etc. in the design process.

5. Embrace diversity in your team. A diverse group can
    help you identify biases early and often.

6. When designing product communication, consider
    any gender, racial, or other biases that might
    unintentionally creep in. For example, show diverse
    options when presenting a search result for a
    doctor or CEO, or provide diverse voice options in a
    personal assistant.

AI systems should treat all stakeholders fairly. In an effort to teach
machines to play fair, we might learn something about ourselves. Some
matters are too important to be left to an unaccountable algorithm.

Product teams should consider which tasks should be delegated to
an AI and which should be performed by people. Bias is often subtle and
unintentional and can creep in without notice. Algorithmic bias can be
extremely harmful to those it affects. Be mindful and intentional about
reducing bias in your AI's results. A biased AI is not trustworthy.

Privacy and Data Collection

Privacy is a key pillar of all AI initiatives.

Data is essential for AI to make informed decisions about people. Any
AI product relies extensively on data to train and improve the system. In
most cases, this data is collected from various public and private datasets,
as well as from the users of the system. People will not share data about
themselves unless they are confident that their privacy is protected and

their data secured.[22] Your AI system should respect the privacy of your users. Privacy and trust are interrelated.

Many near-term AI policy and regulatory issues will focus on the collection and use of data. The development of more effective AI services requires the use of data—often as much relevant data as possible.[23] Your users need to be aware of how their data is collected, used, shared, and stored. Pew Research recently found that being in control of their own information is "very important" to 74% of Americans. The European Commission found that 71% of EU citizens find it unacceptable for companies to share information about them without their permission.[24] Privacy is a key pillar of all AI initiatives.

> *Your users need to be aware of how their data is collected, used, shared, and stored.*

Eventually, the desire for services wins out over a desire for privacy.[25] We like sharing and connecting with people on social media. We prefer getting recommendations on books, music, or movies from Amazon, Spotify, or Netflix. Google Maps is almost essential when navigating. Most users are much less concerned about privacy than we imagine. However, this makes it even more important for you and your team to advocate for users' privacy and empower them with control over their data. The following are some privacy considerations when collecting user data:

1. Protect personally identifiable information.

2. Protect user data.

3. Ask permissions.

---

[22] Smith, Brad, and Harry Shum. *The Future Computed*. Microsoft Corporation, 2018.

[23] Smith, Brad, and Harry Shum. *The Future Computed*. Microsoft Corporation, 2018.

[24] "IBM Design for AI." ibm.com, www.ibm.com/design/ai/.

[25] Kasparov, Garry, and Mig Greengard. *Deep Thinking*. John Murray, 2018.

4. Explain data use.

5. Allow opting out.

6. Consider regulations.

7. Go beyond "terms and conditions."

Protect Personally Identifiable Information

Personally identifiable information or PII is the type of information that can be used to identify a person that can breach their individual privacy. Your social security number, address, phone number, email address, and credit card or bank account details are all types of PII. Exposing PII to bad actors can have severe consequences and can significantly erode trust in your product. Protecting PII is extremely critical. Most products that consume user information use techniques to anonymize this data. The following are two common approaches for anonymizing data:

1. **Aggregation**

    Aggregation is the process of replacing numbers with a summary value. For example, you may replace a list of a user's maximum heartbeats per minute from every day of the month with a single value: their average beats per minute or a categorical high/medium/low label.[26]

2. **Redaction**

    Redaction removes some values to create a less complete picture, for example, an email address is shown as ak****@***il.com. This reduces the probability of identifying users easily.

---

[26] pair.withgoogle.com, https://pair.withgoogle.com/.

However, there is always a possibility that your data contains personally identifiable information. If you're unsure, consult an expert.

Protect User Data

Protect user data from theft, misuse, or data corruption. You should ensure that individual privacy is protected and sensitive and proprietary information is safeguarded. Employ security practices including encryption, access control methodologies, and proprietary consent management modules to restrict access to authorized users and de-identify data according to user preferences. It is your responsibility to work with your team to address any lack of these practices.[27]

Ask Permissions

Ask for appropriate permissions when accessing and collecting user data. Your privacy settings and permissions should be clear, findable, and adjustable.

Explain Data Use

People will be wary of sharing data about themselves if they are not confident of how their data is used, stored, and protected. Provide users with transparency on how their data is collected, used, shared, and stored. Articulate sources of data clearly. Sometimes users can be surprised by their own information when they see it in a new context. These moments often occur when someone sees their data used in a way that appears as if it weren't private or when they see data they didn't know the system had access to, both of which can erode trust.[28] For example, a service that uses a person's financial data to predict loan eligibility without explaining

---

[27] "IBM Design for AI." ibm.com, `www.ibm.com/design/ai/`.
[28] pair.withgoogle.com, `https://pair.withgoogle.com/`.

how it has access to this data is not trustworthy. Communicate privacy and security settings on user data. Explicitly share which data is shared and which data isn't.[29] For example, a social music streaming service that shares what you are listening to with your friends needs to communicate explicitly that your friends can see your listening activity.

Allow Opting Out

Your system should allow users to opt out of sharing certain aspects of their data and sometimes all of their data. Users should be able to deny access to personal data that they may find compromising or unfit for an AI to know or use. Opting out of sharing their entire data can be equivalent to a reset. Consider allowing users to turn off a feature and respect the user's decision to not use it. However, keep in mind that they might decide to use it later and make sure switching it on is easy. While users may not be able to use the AI to perform a task when they have opted out, consider providing a manual, nonautomated way to complete it.

Consider Regulations

Many privacy regulations focus on access and use of data. Consider local and global regulations, and adhere to applicable national and international rights laws when designing your AI's data access permissions.

Go Beyond "Terms and Conditions"

Make your privacy policies understandable. While it is a standard practice to ask users to accept terms and conditions, most users accept them without reading. Digital consent is a misnomer. Terms and conditions or privacy policies are largely designed to provide legally accurate information regarding the usage of people's data to safeguard institutional

---

[29] pair.withgoogle.com, https://pair.withgoogle.com/.

and corporate interests, while often neglecting the needs of the people whose data they process.[30] Most of these documents are inscrutable, they lead to "consent fatigue," and the constant request for agreement to sets of long and unreadable data handling conditions causes a majority of users to simply click and accept terms in order to access the services they wish to use.[31] To build and reinforce trust, go beyond long and unreadable privacy policies. Explain to users how their data is collected, used, shared, and stored in an understandable human language. Make an effort to simplify privacy policies and terms of use whenever possible.

Manipulation

Your AI systems will increasingly extract information from what people are saying and doing through text, voice, video, and other sensors. Your AI might handle sensitive data about people; it might share this data with third parties or itself acquire this data from external sources. Organizations can use this information to build an almost accurate picture or model of an individual or a population. This model can then be used to surface recommendations, actions, products, and services. Sometimes, you might use it to influence people, for example, targeted ads. But you should be aware that this type of manipulation is increasingly dangerous because humans make so many mistakes of the kind that computers are perfectly designed to exploit.[32] AI vs. human games can be really asymmetrical where the AI almost always has the upper hand. Your AI can get very good at finding complex tactics to exploit and manipulate people without their knowledge.

---

[30] Ethically Aligned Design: A Vision for Prioritizing Human Well-Being with Autonomous and Intelligent Systems. First ed., IEEE, 2019.

[31] Ethically Aligned Design: A Vision for Prioritizing Human Well-Being with Autonomous and Intelligent Systems. First ed., IEEE, 2019.

[32] Kasparov, Garry, and Mig Greengard. *Deep Thinking*. John Murray, 2018.

Behavior Control

Sometimes behavior change can be positive, provided that people give their consent to be manipulated. People who want to exercise regularly might not mind being manipulated if the AI helps them and they agree to the behavior change. However, the biggest risk is the power of such an AI system that can control and manipulate people's behaviors in the hands of bad actors. A third person changing your behavior in subtle ways without your knowledge is dangerous. They can use methods of rewards and punishments to directly manipulate individuals or entire populations.

We are extremely vulnerable to misinformation. Content selection algorithms on social media can have insidious effects. Anyone with enough power and savvy can impart false information and circulate fake news. A combination of AI, computer graphics, and speech synthesis makes it possible to generate deepfakes—realistic video and audio content of just about anyone saying or doing just about anything. The technology will require little more than a verbal description of the desired event, making it usable by more or less anyone in the world.[33] Several online marketplaces depend on reputation systems to build trust between buyers and sellers. AI can be used to generate armies of bots and fake accounts that can pump an overwhelming amount of tweets, comments, and recommendations to corrupt online communities and marketplaces where people are merely trying to exchange truthful information. Online misinformation is a matter of public safety with real consequences on real people.

AI can be dangerous when used for the wrong purposes. Be mindful of how your AI affects users, stakeholders, your neighbors, and society at large. Understand the tradeoffs of building AI, ask for the appropriate consent, and give people a choice.

---

[33] Russell, Stuart. *Human Compatible*. Allen Lane, an imprint of Penguin Books, 2019.

Personality

We tend to anthropomorphize, that is, attribute human-like characteristics to, AI systems. Many popular depictions of AI in movies like Samantha in the movie *Her* or Ava in *Ex Machina* show a personality and sometimes even display emotions. AI assistants like Alexa or Siri are designed with a personality in mind. Intentionally imbuing your AI with personality has its advantages and disadvantages. A human-like AI can appear more trustworthy, but your users might overtrust the system or expose sensitive information because they think they are talking to a human. It can set unrealistic expectations about what the AI can and can't do. It can be difficult for people to turn off AI systems if they form an emotional bond with them.

> *A human-like AI can appear more trustworthy, but your users might overtrust the system or expose sensitive information because they think they are talking to a human.*

We've discussed how you can design a good AI personality in Chapter 4 on building trust. Generally, it would be preferable to not imbue your AI with human-like qualities, especially if it is meant to act as a tool like translating languages or recognizing objects from images. However, if you must design your AI systems with a personality, the following are some ethical considerations:

1. **Don't pretend to be human.**

    Your users should know that they are interacting with an AI and not a human. When users confuse an AI with a human being, they can sometimes disclose more information than they would otherwise or rely on the system more than they should.[34] An ethical AI user experience should not sacrifice transparency to appear human-like.

---

[34] pair.withgoogle.com, https://pair.withgoogle.com/.

2. **Clearly communicate boundaries.**

A human-like AI might appear to have human-like capabilities. People can struggle to build accurate mental models of what it can and can't do. Clearly communicate your AI's boundaries. While the idea of a general AI that can answer any questions can be easy to grasp and more inviting, it can set the wrong expectations and lead to mistrust. When users can't accurately map the system's abilities, they may overtrust the system at the wrong times or miss out on the greatest value-add of all: better ways to do a task they take for granted.[35]

3. **Strive for inclusivity.**

Ensure that your AI's personality is inclusive. While you may not be in the business of teaching users how to behave, it is good to establish certain morals for your AI's personality. Consider if you should assign it a gender. Don't perpetuate bad behavior, and consider if your AI marginalizes any person or community. Consider the cultural and societal implications of deploying your AI. Be mindful and intentional about designing your AI's responses.

Risks of Personification

Sometimes a human-like AI can feel more trustworthy; anthropomorphizing AI comes with its own risks. The following are some risks you need to be mindful of:

---

[35] pair.withgoogle.com, https://pair.withgoogle.com/.

1. Think twice before allowing your AI to take over interpersonal services. You need to ensure that your AI's behavior doesn't cross legal or ethical bounds. A human-like AI can appear to act as a trusted friend ready with sage or calming advice but might also be used to manipulate users. Should an AI system be used to nudge a user for the user's benefit or the organization building it?

2. Affective systems make inferences about people's emotions and feelings. When affective systems are deployed across cultures, they could adversely affect the community's cultural, social, or religious values in which they interact.[36]

3. AI personas can perpetuate or contribute to negative stereotypes and gender or racial inequality, for example, suggesting that an engineer is male and a school teacher is female.

4. Human-like AI systems might engage in psychological manipulation of users without their consent.

5. Anthropomorphized AI systems can have side effects such as interfering with the relationship dynamics between human partners and causing attachments between the user and the AI that are distinct from human partnership.[37]

Creating a safe and robust AI is a shared responsibility.

---

[36] Ethically Aligned Design: A Vision for Prioritizing Human Well-Being with Autonomous and Intelligent Systems. First ed., IEEE, 2019.

[37] Ethically Aligned Design: A Vision for Prioritizing Human Well-Being with Autonomous and Intelligent Systems. First ed., IEEE, 2019.

Safe AI

AI systems are not perfect and will sometimes make mistakes. There will also be cases where the AI can be manipulated and hacked for nefarious purposes. AI systems will make a lot of critical decisions for people. Your users' trust in the AI system will depend on whether they can operate the product reliably and safely.

It is essential to ensure the security and robustness of AI systems. While you can run a successful demonstration for a short time, you need an AI system to run for decades with no significant failures in order to qualify as safe.[38] This can mean that you test the AI not only under normal circumstances but also in unexpected conditions or when they are under attack.[39] You need to rigorously test your AI during the development and deployment of your product. Securing AI systems will require developers to identify abnormal behaviors and prevent manipulation, such as the introduction of malicious data that may be intended to negatively impact the AI's behavior.[40]

The principles of robust software design also apply in the case of AI. Engage and involve users, industry participants, governments, academics, and other experts in your design process. The safety and reliability of AI systems will be increasingly important as AI systems become more widely used in fields such as transportation, healthcare, and financial services. For example, Volvo made a competitive advantage out of building safe cars. Ensuring that your AI systems are robust and safe can also give you an edge against the competition.

---

[38] Ford, Martin R. *Architects of Intelligence*. Packt, 2018.

[39] Smith, Brad, and Harry Shum. *The Future Computed*. Microsoft Corporation, 2018.

[40] Smith, Brad, and Harry Shum. *The Future Computed*. Microsoft Corporation, 2018.

Creating a safe and robust AI is a shared responsibility. The following are considerations to promote the safety and reliability of AI systems:

1. Regularly evaluate the quality and suitability of data and models used to train the AI. Share any issues or concerns with your team.

2. Provide appropriate explanations to users about how the AI model works, when it makes mistakes, and the data it uses. This can help users ascribe the right level of trust in the AI when making decisions.

3. Involve domain experts in the design process and operation of AI systems used to make consequential decisions about people.[41]

4. Evaluate when and how the AI should hand off to humans, especially in critical situations like driving or operating heavy equipment.

5. Build robust feedback mechanisms in your product.

Security

AI systems are not foolproof. Researchers have found that it can be surprisingly easy to trick deep learning systems to make errors.[42] The following are some examples:

1. Researchers from Google along with some universities discovered that they could take an ImageNet photo that an AI model predicted correctly with high confidence (e.g., "school bus")

---

[41] Smith, Brad, and Harry Shum. *The Future Computed.* Microsoft Corporation, 2018.

[42] Mitchell, Melanie. *Artificial Intelligence.* First ed., Farrar, Straus and Giroux, 2019.

and distort it by making small, specific changes to its pixels so that the resulting image looked completely unchanged to humans but was now classified with very high confidence by the same AI as something completely different (e.g., "ostrich").[43]

2. One group of researchers created eyeglass frames with specific patterns that could fool facial recognition systems to incorrectly recognize a person.[44]

3. Another group developed stickers that when placed in traffic signs can misclassify the sign.[45]

4. Some researchers showed that an attack on a medical image analysis system can cause a network to change its classification from, say, 99% confidence that the image shows no cancer to 99% confidence that cancer is present.[46] This might be used to create fraudulent diagnostics to charge insurance companies for additional unneeded tests.

---

[43] Elsayed, Gamaleldin F., et al. "Adversarial Examples That Fool Both Computer Vision and Time-Limited Humans." NIPS'18: Proceedings of the 32nd International Conference on Neural Information Processing Systems, 2018, https://doi.org/10.5555/3327144.3327306. Accessed November 25, 2021.

[44] Mitchell, Melanie. *Artificial Intelligence*. First ed., Farrar, Straus and Giroux, 2019.

[45] Mitchell, Melanie. *Artificial Intelligence*. First ed., Farrar, Straus and Giroux, 2019.

[46] Mitchell, Melanie. *Artificial Intelligence*. First ed., Farrar, Straus and Giroux, 2019.

5. Researchers from the University of California at
    Berkeley showed a method to take a sound wave
    and modify it in such a way that the sounds appear
    unchanged to humans, but could be used to fire a
    command to a voice assistant inconspicuously.[47]
    Imagine a hacker broadcasting sounds to your
    Amazon Alexa to start recording and send it to an
    unknown database.

6. NLP researchers have also demonstrated the
    possibility of attacks on sentiment classification
    and question-answering systems. These attacks
    change a few words or add a sentence to a text. It
    does not affect the meaning of the text for a human
    reader, but it causes the system to give an incorrect
    answer.[48]

---

[47] Carlini, Nicholas, and David Wagner. "Audio Adversarial Examples: Targeted Attacks on Speech-To-Text." IEEE Security and Privacy Workshops (SPW), 2018, http://people.eecs.berkeley.edu/~daw/papers/audio-dls18.pdf.

[48] Mitchell, Melanie. *Artificial Intelligence*. First ed., Farrar, Straus and Giroux, 2019.

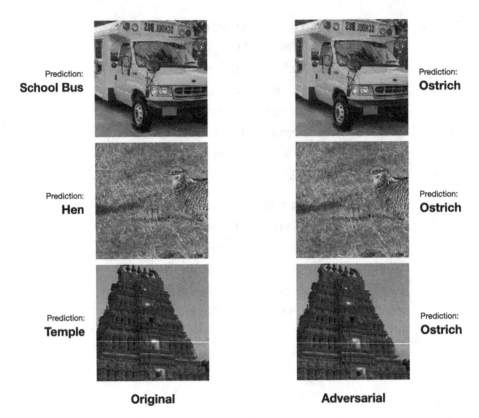

Prediction: **School Bus**

Prediction: **Ostrich**

Prediction: **Hen**

Prediction: **Ostrich**

Prediction: **Temple**

Prediction: **Ostrich**

Original

Adversarial

Figure 7-4. Examples of original and "adversarial" images. The researchers slightly modified the images on the right by manipulating pixels to fool the algorithm into predicting that they are images of ostriches. Source: Szegedy, Christian; Zaremba, Wojciech; Sutskever, Ilya; Bruna, Joan; Erhan, Dumitru; Goodfellow, Ian; Fergus, Rob. Intriguing properties of neural networks. 2013

The ultimate problem is one of understanding. The preceding examples make many researchers question, "What did the AI learn exactly?" Humans wouldn't make these types of mistakes. When we look at a photo of a cat in a house, we can generally gauge its size, shape, color, etc. We can imagine what it could be like to pet it, the sensory feeling of cat fur. Maybe we can also imagine what the cat would do when you bring it treats. AI systems don't have this type of understanding and imagination yet; they can fail in unexpected ways. Most can only look at text in data or the pixels in an image to make inferences. If we mess with the pixels, we can mess with its inference. This is why most critical AI applications need to have a human in the loop. Imagine a self-driving car that starts misclassifying traffic signs. Without human supervision, such systems can be dangerous.

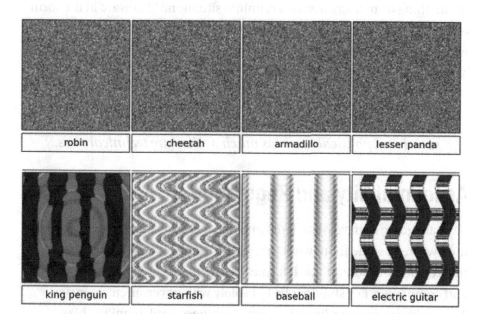

Figure 7-5. Examples of images created specifically to fool AI algorithms. These images are classified with greater than 99.6% confidence by the AI model. Source: `http://karpathy.github.io/2015/03/30/breaking-convnets/`

The methods of fooling AI systems were developed by "white hat practitioners"; these are researchers who actively develop such attacks, look for vulnerabilities, and publish them to help spread awareness and develop defenses. On the other hand, the "black hat" attackers—hackers who are actually trying to fool deployed systems for nefarious purposes—don't publish the tricks they have come up with, so there might be many additional kinds of vulnerabilities of these systems of which we're not yet aware.[49] While researchers have found solutions for specific kinds of attacks, there is still no general defense method.[50]

Detecting unwanted use of AI is another problem. It can be very hard to even detect when there's malicious use of an AI system—which could be anything from a terrorist to a criminal situation.[51] Malware in the form of highly intelligent programs would be much harder to defeat.[52] You might need an AI to defeat such a program. Understanding and protecting against such attacks is a major area of research.

The problems surrounding AI—trustworthiness, explainability, bias, vulnerability to attack, and morality of use—are social and political issues as much as they are technical ones.[53]

Accountability and Regulation

If we are going to have AI systems making decisions that impact people, then we need some framework to regulate them. However, I believe that regulation shouldn't be left solely at the hands of AI researchers and companies, nor should it be left solely with government agencies. The problems surrounding AI—trustworthiness, explainability, bias, vulnerability to attack, and morality of use—are social and political issues as much as they are technical ones.[54] It is crucial that the discussion around these issues include people with different perspectives and backgrounds.

---

[49] Mitchell, Melanie. *Artificial Intelligence*. First ed., Farrar, Straus and Giroux, 2019.

[50] Mitchell, Melanie. *Artificial Intelligence*. First ed., Farrar, Straus and Giroux, 2019.

[51] Ford, Martin R. *Architects of Intelligence*. Packt, 2018.

[52] Russell, Stuart. *Human Compatible*. Allen Lane, an imprint of Penguin Books, 2019.

[53] Mitchell, Melanie. *Artificial Intelligence*. First ed., Farrar, Straus and Giroux, 2019.

[54] Mitchell, Melanie. *Artificial Intelligence*. First ed., Farrar, Straus and Giroux, 2019.

However, regulation shouldn't start with the view that its goal is to stop AI or hold back the deployment of these technologies.[55] Some researchers believe that regulation should be modeled after regulation of other fields, particularly genetic engineering and medical sciences. For example, the Asilomar Conference on Biotechnology Ethics publishes and updates ethical standards and strategies on a regular basis. These standards have worked very well, and the number of people who have been harmed by intentional or accidental abuse or problems with biotechnology has been close to zero.[56] Many of these comprehensive ethical standards and technical strategies on how to keep the technology safe have now been baked into the law.

Accountability

A looming question that comes up when AI systems start making decisions on behalf of people is one of accountability. Who should be responsible when the AI fails or harms others? Is it the company that built the AI, the team that designed the system, the organization that bought or decided to use it, or the person who allowed the AI to make the harmful decision? These are hard questions that product teams need to consider when deploying AI. Many believe that the people who design and deploy AI systems must be accountable for how their systems operate.[57] Every person involved in the creation of AI at any step is accountable for considering the system's impact in the world, as are the companies invested in its development.[58] Here are some steps to ensure accountability[59]:

---

[55] Ford, Martin R. *Architects of Intelligence*. Packt, 2018.

[56] Ford, Martin R. *Architects of Intelligence*. Packt, 2018.

[57] Smith, Brad, and Harry Shum. *The Future Computed*. Microsoft Corporation, 2018.

[58] "IBM Design for AI." ibm.com, www.ibm.com/design/ai/.

[59] "IBM Design for AI." ibm.com, www.ibm.com/design/ai/.

1. Make company policies clear and accessible to design and development teams from day one so that no one is confused about issues of responsibility or accountability.

2. Understand where the responsibility of your AI ends. You might not have control over how certain users choose to employ the AI.

3. Keep detailed records of your design processes and decision-making. Determine a strategy for keeping records during the design and development process to encourage best practices and encourage iteration.

4. Adhere to your company's guidelines about business conduct. Understand national and international laws, regulations, and guidelines that your AI may have to work within.

Law

Your team will need to comply with a broad range of international and local laws governing fairness, privacy, injuries resulting from unreasonable behaviors, and the like. However, we also need to develop clear principles to guide people to build, use, and apply AI systems. Industry groups and others should build off these principles to create detailed best practices for key aspects of the development of AI systems, such as the nature of the data used to train AI systems, the analytical techniques deployed, and how the results of AI systems are explained to people using those systems.[60]

---

[60] Smith, Brad, and Harry Shum. *The Future Computed*. Microsoft Corporation, 2018.

Liability

It is essential to clarify who is legally responsible for actions by automated systems. This can depend on the nature and type of application. You might have different legal frameworks for shopping assistants vs. self-driving cars. If an agent makes a monetary transaction on behalf of a user, is the agent or the user liable for any debts that arise? Would it make sense for the AI agent to have their own assets separate from the user? So far, these questions are not well-understood.[61]

According to some researchers, well-tested principles of negligence law are most appropriate for addressing injuries arising from the deployment and use of AI systems.[62] Negligence laws hold parties accountable if they fall short of the promised standard and encourage good conduct.

Independent Review Committees

You can think of forming independent review committees of experts from diverse fields to oversee whether the AI system meets ethical and legal criteria during development and after it is deployed. According to the IEEE Ethically Aligned Design document, an independent, internationally coordinated body—akin to ISO—should be formed, which should also include a certification process.[63]

---

[61] Russell, Stuart J., and Peter Norvig. *Artificial Intelligence.* Third ed., Pearson, 2016.

[62] Smith, Brad, and Harry Shum. *The Future Computed.* Microsoft Corporation, 2018.

[63] Ethically Aligned Design: A Vision for Prioritizing Human Well-Being with Autonomous and Intelligent Systems. First ed., IEEE, 2019.

There is also a possibility where your AI system becomes so accurate that it might be legally required to use its recommendation. For example, if expert systems become reliably more accurate than human diagnosticians, doctors might become legally liable if they don't use the recommendations of these expert systems.[64]

Beneficial AI

Because AI systems will make predictions and decisions for people and on behalf of them, they need to work for you and me. Our AI systems need to be beneficial to us, that is, we can expect them to achieve the objectives that we want. We need to make sure that people are provably better off with the AI system than without.

> *The goals we give the AI might not always align with what we actually want.*

Control Problem

The simplest way to communicate an objective to an AI is in the form of goals,[65] that is, it either achieves the goal or not. In the standard model used in most AI systems, we put a purpose into the machine by giving it a goal. The problem is not that we might fail to do a good job of building AI systems; it's that we might succeed too well.[66] The goals we give the AI might not always align with what we actually want. Nick Bostrom, in his book *Superintelligence*, presented a thought experiment in which an

---

[64] Russell, Stuart J., and Peter Norvig. *Artificial Intelligence*. Third ed., Pearson, 2016.

[65] Russell, Stuart. *Human Compatible*. Allen Lane, an imprint of Penguin Books, 2019.

[66] Russell, Stuart. *Human Compatible*. Allen Lane, an imprint of Penguin Books, 2019.

AI, designed to manage production in a factory, is given the final goal of maximizing the manufacture of paperclips and proceeds by converting first the earth and then increasingly large chunks of the observable universe into paperclips.[67] This is the control problem. We may suffer failure of value alignment, that is, we might assign the AI objectives that might be imperfectly aligned with our own. No one wants the universe to turn into paperclips.

Beneficial Machines

A beneficial machine is whose actions can be expected to achieve our objectives rather than its objectives.[68] We should not be afraid of intelligent machines, but of machines making decisions that they do not have the intelligence to make.[69] The economist Sendhil Mullainathan highlighted the notion of tail risk. He wrote, "I am far more afraid of machine stupidity than of machine intelligence. Machine stupidity creates a tail risk. Machines can make many, many good decisions and then one day fail spectacularly on a tail event that did not appear in their training data."[70]

> *A beneficial machine is whose actions can be expected to achieve our objectives rather than its objectives.*[71]

We don't necessarily want machines with our type of intelligence. We actually want machines whose actions can be expected to achieve our objectives, not their objectives.[72] We need to define AI in such a way

---

[67] Bostrom, Nick. *Superintelligence.* Oxford University Press, 2016.

[68] Russell, Stuart. *Human Compatible.* Allen Lane, an imprint of Penguin Books, 2019.

[69] Mitchell, Melanie. *Artificial Intelligence.* First ed., Farrar, Straus and Giroux, 2019.

[70] Mitchell, Melanie. *Artificial Intelligence.* First ed., Farrar, Straus and Giroux, 2019.

[71] Russell, Stuart. *Human Compatible.* Allen Lane, an imprint of Penguin Books, 2019.

[72] Ford, Martin R. *Architects of Intelligence.* Packt, 2018.

that it remains under the control of the humans that it's supposed to be working for.[73]

Principles of Beneficial Machines

In his book *Human Compatible,* Stuart Russel has outlined some principles for AI researchers and developers to think about how to create beneficial AI systems. These are meant to be guidelines and not explicit instructions. The following are the three principles[74]:

1. The machine's only objective is to maximize the realization of human preferences.

2. The machine is initially uncertain about what those preferences are.

3. The ultimate source of information about human preferences is human behavior.

The first principle that "the machine's only objective is to maximize the realization of human preferences" is central to the idea of beneficial machines. Preferences cover anything you might care about now or in the future. The machine is purely altruistic—that is, it attaches absolutely no intrinsic value to its own well-being or even its own existence.[75]

Humans don't have any preferences in a meaningful sense. We aren't born with preferences, and our priorities change over time. The second principle that "the machine is initially uncertain about what those preferences are" is essential for handling evolving preferences. A machine with a single-minded goal of achieving an objective will not ask

---

[73] Ford, Martin R. *Architects of Intelligence.* Packt, 2018.

[74] Russell, Stuart. *Human Compatible.* Allen Lane, an imprint of Penguin Books, 2019.

[75] Russell, Stuart. *Human Compatible.* Allen Lane, an imprint of Penguin Books, 2019.

if a particular mathematically optimal course of action is desirable. It will ignore humans even when they object. Assuming perfect knowledge of the objective decouples the machine from the human: what the human does no longer matters, because the machine knows the goal and pursues it.[76] It might even stop humans from switching it off. On the other hand, a machine that is uncertain will exhibit humility. It will defer to humans when it is unsure and allow itself to be switched off.

The third principle that "the ultimate source of information about human preferences is human behavior" is based on the idea that human choices reveal information about their preferences. As human preferences evolve, the machine can observe and change accordingly. The AI becomes more beneficial as it learns from preferences. Humans are not perfectly rational: imperfection comes between human preferences and human choices, and the machine must take into account those imperfections if it is to interpret human choices as evidence of human preferences.[77] This is an ongoing area of research.

Human in the Loop

Till we build perfectly beneficial machines, AI will still make some mistakes. When we say that an AI is 95% accurate, five out of a hundred results will be incorrect. You can trust such a system for low-stakes situations like movie recommendations, but you would be wary of trusting it for conducting surgery or diagnosing cancer. All AI systems are probabilistic and are therefore error-prone. They are bound to make mistakes at some point.

---

[76] Russell, Stuart. *Human Compatible*. Allen Lane, an imprint of Penguin Books, 2019.

[77] Russell, Stuart. *Human Compatible*. Allen Lane, an imprint of Penguin Books, 2019.

While an AI system is excellent at churning data, finding patterns, or detecting anomalies, it has no understanding of what it is doing. Humans are better at handling uncertain and unpredictable situations. Take the example of a judge deciding a person's fate—sentence them to jail or free them. Criminal judgment is often a difficult, moral decision. Even if an AI judge is 99% accurate, this means that one person out of a hundred is incorrectly sentenced, which is cruel. Most of us would still prefer an imperfect human being as a judge rather than an AI that doesn't understand what it is doing.

While it might be easy to focus on these shortcomings, AI has its advantages in improving the speed and quality of outcomes. Sometimes these improvements are superhuman. There are huge benefits that AI systems already bring to society, and they can be even more beneficial. Many of our most pressing problems like climate change, food safety, drug discovery, or monitoring health and safety might be extremely complex. According to Demis Hassabis, the founder of DeepMind, it would be difficult for individual humans and scientific experts to have the time they need in their lifetimes to even innovate and advance.[78] We will need some help from AI.

In many cases, it would make sense to use AI while accounting for a margin of error. To account for errors, we need ways for people to oversee the system and take control if things go wrong. We need humans in the loop.

Summary

We discussed the role of AI ethics in designing trustworthy, safe, and beneficial AI experiences. The following are some key points:

---

[78] Mitchell, Melanie. *Artificial Intelligence*. First ed., Farrar, Straus and Giroux, 2019.

1. AI is used in critical and varied applications, from recommending jobs, criminal sentencing, and airplane navigation systems to cybersecurity. If we are going to let AI make decisions, it should be aligned with the ethics of the society in which it operates.

2. AI ethics is a system of principles, values, and techniques intended to inform artificial intelligence development and responsible use. AI ethics can be about the behavior or principles of teams that design AI systems or the behavior of the AI itself.

3. Raising ethical concerns can feel like an impediment to a project. Ethics can take a back seat, especially in fast-growing industries and companies. However, ignoring ethical implications can erode user trust in your AI. An AI that is not trustworthy is not useful.

4. Encourage people to talk about ethical implications if they arise. Try to incentivize team members to raise ethical concerns.

5. AI ethics is a large subject, and it would not be easy to discuss everything from the topic in one chapter. We focused on a subset of ethical considerations relevant for designing AI products:

   a. Ensuring your AI systems are trustworthy

   b. Ensuring your AI is safe

   c. Ensuring your AI systems are beneficial

6. Users need to be able to trust your AI's decisions for them to use the product consistently.

7.   Building trust goes beyond just the user interface. Your sales and marketing communications, R&D efforts, PR mentions, awards, fundraise announcements, and customer testimonials all contribute to reinforcing trust in your AI.

8.   Trust is a critical consideration when building AI products. Ensure that your AI systems are explainable, fair, reliable, and inclusive and respect the privacy of their users and those they affects.

9.   Bias matters. Algorithmic bias is the tendency of machine learning algorithms to produce inappropriately biased decisions. As AI systems are used to make crucial decisions that affect people and societies, we need to ensure that these decisions are fair, do not put certain individuals at a disadvantage, or do not reinforce undesirable stereotypes.

10.   In most cases, bias in AI algorithms is not intentional. It occurs because of faults in the underlying data on which the AI was trained.

11.   Privacy is a key pillar of all AI initiatives. Most users are much less concerned about privacy than we imagine. However, this makes it even more important for you and your team to advocate for users' privacy and empower them with control over their data.

12.   AI can be dangerous when used for the wrong purposes. Be mindful of how your AI affects users, stakeholders, your neighbors, and society at large. Understand the tradeoffs of building AI, ask for the appropriate consent, and give people a choice.

13. Intentionally imbuing your AI with personality has
     its advantages and disadvantages. A human-like AI
     can appear more trustworthy, but your users might
     overtrust the system or expose sensitive information
     because they think they are talking to a human. It
     can set unrealistic expectations about what AI can
     and can't do.

14. AI systems will make a lot of critical decisions
     for people. Your users' trust in the AI system will
     depend on whether they can operate the product
     reliably and safely. It is essential to ensure the
     security and robustness of AI systems.

15. The principles of robust software design also apply
     in the case of AI. Engage and involve users, industry
     participants, governments, academics, and other
     experts in your design process. Ensuring that your
     AI systems are robust and safe can also give you an
     edge against the competition.

16. AI systems are not foolproof. Researchers have
     found that it can be surprisingly easy to trick deep
     learning systems.

17. If we are going to have AI systems making decisions
     that impact people, then we need some framework
     to regulate them. The problems surrounding AI—
     trustworthiness, explainability, bias, vulnerability to
     attack, and morality of use—are social and political
     issues as much as they are technical ones. The
     discussion around these issues must include people
     with different perspectives and backgrounds.

18. Because AI systems will make predictions and decisions for people and on behalf of them, they need to work for you and me. Our AI systems need to be beneficial to us, that is, we can expect them to achieve the objectives that we want. We need to make sure that people are provably better off with the AI system than without.

19. Till we build perfectly beneficial machines, AI will still make some mistakes. To account for errors, we need ways for people to oversee the system and take control if things go wrong. We need humans in the loop.

CHAPTER 8

Prototyping AI Products

Prototyping AI products can be tricky. In this chapter, we discuss the different ways you could prototype AI experiences depending on the stage of your product and what you want to measure.

We have two cats in our house. And anyone who has lived with cats knows that they can be incredibly picky about everything from furniture, toys, food, or even people. Whenever we have guests over, we know that one of them will hide in the farthest corner of the bedroom and won't come out till a few hours have passed. Our cats are especially fussy about their food. There have been times when I've bought a crate of a new type of treat only to realize that they would not go within 10 feet of it. Over the years, we've learned that it is better to try out smaller samples of one or two packets before buying a dozen. In short, we test before we make a large investment.

Building AI products is time-consuming and expensive. Prototyping is a way of trying out things quickly in a low-risk way before making a significant investment of time and resources. It is a common misconception that prototyping happens in a lab and is complex and expensive. When we say prototype, that's anything that gets the idea

A. Kore, *Designing Human-Centric AI Experiences*,
https://doi.org/10.1007/978-1-4842-8088-1_8

in your head into an artifact people can experience and offer feedback on.[1] It could be a sketch on a piece of paper, a mockup, a deck, or even a conversation to role-play what it might feel like to use a product or service. A prototype can help you tell a story, start conversations, or even convince stakeholders. The following are some benefits of prototyping:

1. They can help you demonstrate your ideas and communicate with your team.

2. They can help convince adamant stakeholders.

3. They can start conversations and align the team in a particular direction.

4. They can help you and your team to brainstorm and generate new ideas.

5. They can save time and money in the long run by getting early feedback before you make an investment of resources.

6. They can help your team identify core problems, what to build, as well as what not to build.

Sometimes it can feel like prototyping slows down the product development process, but in the long run, you'll save time and money by getting to the right idea faster.[2]

A prototype is anything that gets the idea in your head into an artifact that people can experience and offer feedback on.

---

[1] "Why Everyone Should Prototype (Not Just Designers)." IDEO U, www.ideou.com/ blogs/inspiration/why-everyone-should-prototype-not-just-designers.

[2] "Why Everyone Should Prototype (Not Just Designers)." IDEO U, www.ideou.com/ blogs/inspiration/why-everyone-should-prototype-not-just-designers.

Prototyping AI Experiences

It can be tricky to prototype AI products, especially if the value of your product lies in using user information to build a personalized experience; a prototype with dummy data wouldn't feel authentic. And if you have a fully built ML system, it might be too late to make any meaningful changes. AI systems also introduce additional challenges like explainability and relevance of AI results, which may not be easy to prototype. So when building a low-risk, low-cost mechanism for testing your AI experience, it is good to understand the prototype's goal.

You might test your prototype against the following objectives:

1. Desirability

2. Usability

3. Explainability

4. Relevance

Desirability

When prototyping to check the desirability of your product, you are testing for motivation. *Desirable* refers to when a product or service is valuable and required and someone will pay for it. If you hypothesize that users will benefit from a new feature or product, this type of prototype can help you understand if it is worthwhile to build. Sometimes a feature might be useful, but your customers won't be willing to pay for it. This test can also help you check if the solution is economically viable.

> Desirable *refers to when a product or service is valuable and required and someone will pay for it.*

A prototype for desirability can be a popup in your application, a fake ad on Facebook or Google, a survey, or simply a conversation with users. You might show them a mockup to gauge interest. This is also known as a "smoke test" or "painted door test." Keep in mind that this type of prototype is not an end-to-end experience. You are only trying to see if customers are interested in using or buying your solution in a quick, low-cost, and low-risk manner.

Figure 8-1. *"Smoke screen" or "painted door" test for a service that analyzes profile pictures. Users are shown a fake experience to gauge interest*

Usability

When most people think of prototypes, they think of a usability prototype. A usability prototype is a draft version of a product that allows you to explore your ideas and show the intention behind a feature or the overall design concept to users before investing time and money into development.[3] It can be anything from a sketch of the interaction, a mockup, or a slideshow to a fully functioning application.

---

[3] "Prototyping | Usability.gov." usability.gov, www.usability.gov/how-to-and-tools/methods/prototyping.html.

A usability prototype is a draft version of a product that allows you to explore your ideas before investing considerable effort.

Most design professionals use the term *fidelity* to describe prototypes. Fidelity refers to the level of detail and closeness to the actual experience. The prototype can be low fidelity like a sketch, mockup, or slideshow, or it can be high fidelity like a detailed end-to-end workflow or a fully functional application. The goal of a usability prototype is to check if the target users can use the product as expected. It helps teams identify points of confusion or failure in the user experience. You might use these types of prototypes to test a small feature or even end-to-end workflows.

Types of Usability Prototypes

If the value of your AI product lies in a contextual and personalized experience, it can be challenging to build usability prototypes. However, there are a few approaches that can help:

1.   Using personal examples

2.   Wizard of Oz studies

3.   Minimum viable product (MVP)

Using Personal Examples

When prototyping with early mockups, you can ask your users to get personal data like contact lists, photos, favorite songs, etc. to the sessions. This can be like a fun homework for participants. With these examples, you can then simulate right and wrong responses from the system. For example, you can simulate the system returning the wrong movie recommendation to the user to see how they react and what assumptions they make about why the system returned that result. This helps you assess the cost and benefits of these possibilities with much more validity than using dummy examples or conceptual descriptions.[4] However, make sure that you inform the participants how the data would be used and if or when it would be deleted.

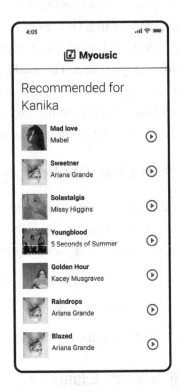

 Generic    Personal examples

Figure 8-2. Examples of a generic (left) and personalized (right) prototype of a music recommendation app

---

[4] Holbrook, Jess. "Human-Centered Machine Learning." Medium, 2017, `https://medium.com/google-design/human-centered-machine-learning-a770d10562cd`.

Wizard of Oz Studies

Wizard of Oz studies have participants interact with what they believe is an autonomous system. For example, when testing a chat-based interface, a teammate can simulate answers from what would be the AI system. Having a teammate imitate actions like answering questions, giving movie recommendations, etc. can simulate "intelligent" systems. When participants engage with what they perceive to be AI, they tend to form mental models and adjust their behavior accordingly. Observing these changes can be extremely valuable when designing AI systems.

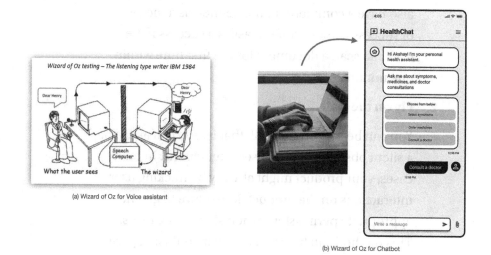

Figure 8-3. Wizard of Oz. (a) Example of a Wizard of Oz test for a voice assistant. Source: www.simpleusability.com/inspiration/2018/08/wizard-of-oz-testing-a-method-of-testing-a-system-that-does-not-yet-exist/. (b) Simulating a chatbot experience. The participant feels like they are interacting with an AI that is simulated by a human remotely. Image source: Photo by Christin Hume on Unsplash

Minimum Viable Product

Sometimes when testing a critical feature, you might decide to build a functional prototype of the experience. This is known as a minimum viable product (MVP). You might give your users access to the MVP and test for usability using some of the common techniques shown in the following:

1. **Task completion**

   You can ask your users to complete a predefined task on your product or a prototype and observe if they are able to accomplish the task. You can also have a completion time defined as a metric of success. For example, a task is a success if the user can scan a document for the first time within one minute.

2. **Fly on the wall**

   You can be a fly on the wall, that is, you can ask to be a silent observer of your user interactions. In many cases, your product might already be tracking user interactions on the product. Make sure to get the appropriate permissions when shadowing users. This method can help you uncover confusing parts of your product, where users hesitate, points of failure, and success.

MVP is not the first version of your product.

Keep in mind that an MVP is not the first version of your product. It is only used for learning and validating assumptions. Building an MVP will require you to engage your machine learning and engineering counterparts. So you need to understand the importance of good enough and make appropriate tradeoffs.

Explainability

While your AI products need to be usable, it is equally important for your AI's explanations to be understandable. When designing for AI explainability, you need to assess if your explanations increase user trust or make it easier for users to make decisions. Explainability prototypes can be in the form of mockups, product copy, surveys, and user interaction prototypes. When embarking on a new project, you can evaluate your explanations internally within your team and later with users.

You need to assess if your explanations increase user trust or make it easier for users to make decisions.

Internal Assessment

You can evaluate your explanations with your product managers, designers, machine learning scientists, and engineers on your team. You can conduct your assessment on the following points:

1. Consider if your type of explanation is suitable for the user and the kind of product.

2. Observe how your team members interact with the explanation. Ask them what they understand from the explanation and what parts are confusing.

3. Determine if the components of your explanation are relevant to the user. Are we highlighting the right parts in the explanation?

4. Determine if your explanation has any implications on user privacy, proprietary information, or the product's security.

User Validation

You should also validate your explanations with the users of your product. Your user group should reflect the diversity of your audience. You can use different methods to prototype your explanations. While these methods can be subjective, they provide great insight into how users perceive your explanations and if they are helpful.

1. **User interviews**

   You can conduct one-on-one interviews with users and ask them what they think about your explanation. You need to check if their mental model matches your product's model of how the AI works.

2. **Surveys and customer feedback**

   You can float survey forms to your customers or ask for feedback inside the product while interacting with it. You might sometimes be able to validate your explanations by listening to customer service calls, reviews and feedback of your product on external websites and app stores, and public conversations about your product on social media.

3. **Task completion**

   You can ask your users to complete a predefined task on your product or a prototype and observe if your explanation helps them accomplish the task. You can also have a completion time defined as a metric of success. For example, an explanation is a success if the user is able to scan a document for the first time within one minute.

4. **Fly on the wall**

You can be a fly on the wall, that is, you can ask to be a silent observer on your user interactions. In many cases, your product might already be tracking user interactions on the product. Make sure to get the appropriate permissions when shadowing users. This method can help you uncover confusing parts of your product, where users hesitate, where you need to explain better, and which parts need explanations.

🚗 In-car navigation

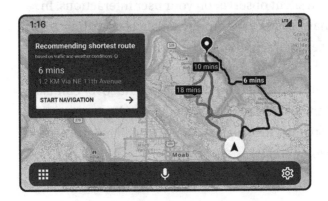

✅ **Aim for**

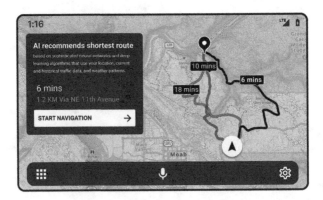

❌ **Avoid**

Figure 8-4. Examples of explanations. (Top) Informing the user that the shortest route is recommended is an easy explanation that most users can understand and act on. (Bottom) In this case, a detailed description of how the AI system works is not useful

Relevance

This type of prototype helps you check the relevance of your AI's results. Building relevance prototypes is complicated and will require you to collaborate with your machine learning and engineering counterparts. A relevance prototype can be high fidelity like an MVP, a Wizard of Oz prototype where you simulate the AI's results, or even something as simple as a tabular view with the list of your model's suggestions. You can show these lists of results to users and get feedback on their relevance.

Hardware Prototypes

Most examples of prototyping discussed in this chapter focus on AI as software. Building physical mockups and prototypes of hardware is much more difficult and time-consuming. To avoid the time and expense of working with physical electronics and mechanics too early in the process, designers can simulate the behavior of a physical AI device in 3D within a game engine like Unity3D.[5] They can then use augmented reality to visualize this simulation in a real-world environment. Once the designer is ready to build a physical prototype, they can use various components like a Raspberry Pi or Arduino board, servo motors, LEDs, speakers, etc.

---

[5] "Prototyping Ways of Prototyping AI | ACM Interactions." interactions.acm.org, 2018, https://interactions.acm.org/archive/view/november-december-2018/prototyping-ways-of-prototyping-ai.

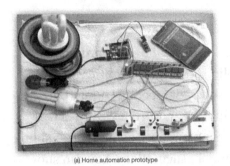

(a) Home automation prototype

(b) Google's self-driving car prototype

Figure 8-5. ***Examples of hardware prototypes.*** *(a) Prototype of an Arduino-based home automation system. Source:* www. electronicsforu.com/electronics-projects/hardware-diy/ arduino-home-automation-system-android. *(b) Google's self-driving car prototype. Source:* www.wired.com/2014/05/google-self-driving-car-prototype/

Summary

We discussed different ways you could prototype AI experiences depending on the stage of your product and what you want to measure. The following are some notable points:

1. Building AI products is time-consuming and expensive. Prototyping is a way of trying out things quickly in a low-risk way before investing time and resources significantly.

2. A prototype is anything that gets the idea in your head into an artifact people can experience and offer feedback on.

3. A prototype can help you tell a story, start conversations, or even convince stakeholders.

4. It is good to understand the prototype's goal. You might test your AI prototype against the following objectives:

   a. Desirability

   b. Usability

   c. Explainability

   d. Relevance

5. *Desirable* refers to when a product or service is valuable and required and someone will pay for it.

6. A usability prototype is a draft version of a product that allows you to explore your ideas and show the intention behind a feature or the overall design concept. The following are a few approaches that can help you build usability prototypes for AI experiences:

   a. Using personal examples

   b. Wizard of Oz

   c. Minimum viable product (MVP)

7. Explainability prototypes can be in the form of mockups, product copy, surveys, and user interaction prototypes. When embarking on a new project, you can evaluate your explanations internally within your team and later with users.

8. A relevance prototype helps you check the relevance of your AI's results.

9. Building physical mockups and prototypes
   of hardware is much more difficult and time-
   consuming. Designers can simulate prototypes in
   a 3D environment and eventually build a physical
   working prototype.

PART 4

Teamwork

CHAPTER 9

Understanding AI Terminology

In this chapter, we will look at some mainstream AI techniques, metrics used to evaluate the success of AI algorithms, and common capabilities of AI.

A large part of what AI teams do is figure out problems to solve with AI and match appropriate techniques and capabilities to the solution. While most product-facing PMs and designers are used to working together, they often speak different languages and have different working styles from their engineering and machine learning counterparts. These differences can inhibit powerful collaborations. Product designers and product managers need to understand what their tech teams do and speak their language to work effectively within AI teams.

In this chapter, we will look at the following:

1. Mainstream AI techniques

2. Metrics used to measure AI performance

3. Common AI capabilities and the applications they enable

A large part of what AI teams do is figure out problems to solve with AI and match appropriate techniques and capabilities to the solution.

Key Approaches for Building AI

There are different ways to implement AI, but at a high level, they are based on two approaches, namely, rule-based and examples-based. In the rule-based approach, the AI is programmed to follow a set of specific instructions. The algorithm tells the computer precisely what steps to take to solve a problem or reach a goal. For example, a robot designed to navigate a warehouse is given specific instructions to turn when there is an obstruction in its path.

In the examples-based approach, the AI is taught, not programmed. Learning happens by giving the AI a set of examples of what it will encounter in the real world and how to respond. For example, an AI may be shown multiple images of faces in different environments to detect faces from an image. By observing these images, the AI learns how to detect a face.

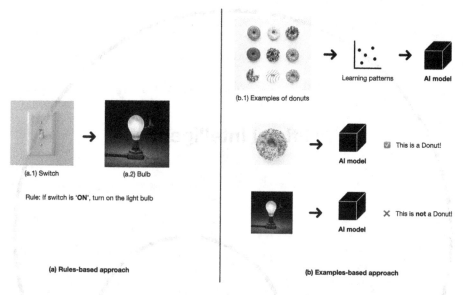

Figure 9-1. Approaches of building AI. *(a) Rules-based approach. Here, the switch is mapped to the light bulb. The rule is to turn on the light bulb by flicking the switch on. (b) Examples-based approach. The AI learns to recognize donuts by showing examples of what a donut looks like without specifying exact steps*

The most popular AI techniques are a combination of rule-based and examples-based approaches. In larger AI products, it is common to use multiple AI techniques to achieve an objective. For example, a smart speaker might use a different AI algorithm to detect a voice and a different one to search for a response.

AI is an evolving field, and many new techniques will get invented over time. A significant amount of progress in AI techniques has been made in the subfield of machine learning (ML), so much so that AI in the industry is sometimes synonymous with ML.

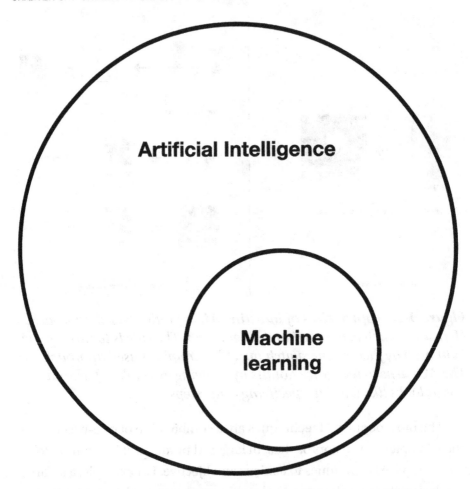

Figure 9-2. *Artificial intelligence is a broad field, and machine learning is a subset of AI*

AI Techniques

This section will look at some of the following mainstream ML techniques used by AI product teams:

1. Supervised learning

2. Unsupervised learning

3. Reinforcement learning

4. Deep learning and neural networks

5. Backpropagation

6. Transfer learning

7. Generative adversarial networks (GANs)

8. Knowledge graphs

Supervised Learning

In supervised learning, you train the AI by mapping inputs to outputs. We use labeled data to build the model. *Labeled* means that along with the example, we also provide the expected output. For instance, if we want to build a system that detects product defects in a factory, we can show the ML model many examples of defective and good-condition products. We can label these examples as "defective" or "good condition" for the model to learn. We can label a face recognition dataset by indicating whether an example image contains a face or not.

This technique is a key part of modern AI, with the most value created by supervised learning algorithms. Supervised learning is a common technique that accounts for about 95% of all practical applications of AI, and it has created a lot of opportunities in almost every major industry.[1]

Most supervised learning focuses on two types of techniques, namely, classification and regression:

1. **Classification** refers to the grouping of the AI's results into different categories. In this technique, the AI learns to group labeled data by categories,

---

[1] Ford, Martin R. *Architects of Intelligence*. Packt, 2018.

for example, grouping photos by people, predicting
whether or not a person will default on a loan,
sorting eggs by quality in an assembly line, etc.

2. **Regression** is a statistical technique that finds
a prediction based on the average of what has
happened in the past.[2] The idea is to take past data
and use it to predict future outcomes like predicting
flight prices or detecting anomalies in data, for
example, credit card fraud, where specific behavior
patterns may be observed.

However, today, to teach a computer what a coffee mug is, we show it
thousands of coffee mugs, but no child's parents, no matter how patient
and loving, ever pointed out thousands of coffee mugs to their child.[3] The
challenge with supervised learning is that it needs massive amounts of
labeled data. Getting accurate and useful labeled data can be costly and
time-consuming. Companies often employ human annotators to label
training data.

---

[2] Agrawal, Ajay, et al. *Prediction Machines*. Harvard Business Review Press, 2018.

[3] Ng, Andrew. AI for Everyone. `www.coursera.org/learn/ai-for-everyone`.

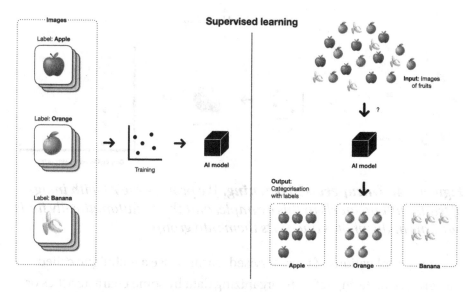

*Figure 9-3. **Supervised learning.** In this example, we train the AI with labeled data of images of fruits. After training, the AI model can categorize fruit images by their labels*

Unsupervised Learning

Most people would agree that if you have to teach a child to recognize cats, you don't need to show them 15,000 images to understand the concept. In unsupervised learning, there is no labeled data. We teach machines to learn directly from unlabeled data coming from their environments. This is somewhat similar to how human beings learn. With unsupervised learning, you essentially give the algorithm lots of data and ask it to find patterns in that data. The AI learns by finding patterns in the input data on its own.

Unsupervised learning

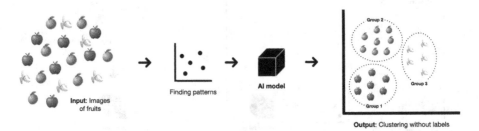

Figure 9-4. Unsupervised learning. *We provide the AI with images of fruits with no labels in this example, and the AI automatically finds patterns in the photos and sorts them into groups*

Many applications of unsupervised learning use a technique called clustering. Clustering refers to organizing data by some characteristics or features. These features are then used as labels that the model generates. This technique is especially useful when you want to group things like a set of images by people, animals, landscapes, etc. or find groups of similar users for targeting ads.

Figure 9-5. Clustering. *A pile of vegetables is clustered by its type in this example. Source:* `https://medium.com/mlearning-ai/cluster-analysis-6757d6c6acc9`

Access to labeled data is often a bottleneck for many AI projects, and this is also one of the most difficult challenges facing the field. Unsupervised learning is an extremely valuable technique and represents one of the most promising directions for progress in AI. With unsupervised learning, we can imagine systems that can learn by themselves without the need for vast volumes of labeled training data.

Reinforcement Learning

Sometimes it can be difficult to specify the optimal way of doing something. In reinforcement learning, the AI learns by trial and error. In this technique, AI is rewarded for positive behavior and punished for negative behavior.

We show the AI data, and whenever it produces correct output, we reward it (sometimes in the form of scores). When it produces incorrect results, we punish it by not rewarding it or deducting rewards. Over time the AI builds a model to get the maximum reward.

Training a pet is a good analogy for understanding reinforcement learning. When the pet displays good behavior, you give it treats, and when not, you don't give it any treats. Over time the pet learns that if they behave in a particular manner, they will get rewards, and you get the pet's good behavior as an outcome.

Reinforcement learning

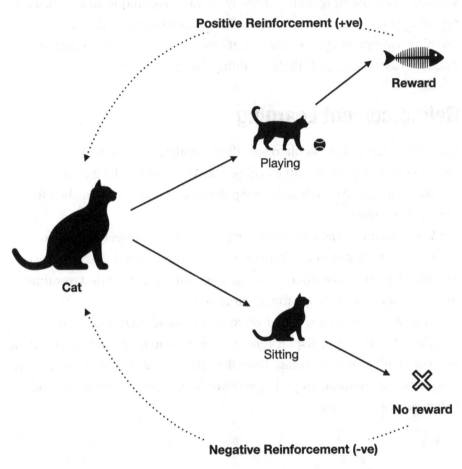

Figure 9-6. *Training a cat using positive and negative reinforcement*

Reinforcement learning is used for building AI that drives autonomous vehicles like self-driving cars, drones, etc., as well as for playing games like chess, *Go*, and video games. The AI can play a lot of games against itself to learn on its own. Google's DeepMind team trained their AI to play 49 different Atari video games entirely from scratch, including *Pong*, *Freeway*, and *Space Invaders*. It used only the screen pixels as input and the game score as a reward signal.[4]

The main challenge with reinforcement learning is that it requires a large number of practice runs and massive computing resources before the algorithm is usable in the real world. A human can learn to drive a car in 15 hours of training without crashing into anything. If you want to use the current reinforcement learning methods to train a car to drive itself, the machine will have to drive off cliffs 10,000 times before it figures out how not to do that.[5]

Deep Learning and Neural Networks

Deep learning (DL) is a subset of machine learning. In ML, learning happens by building a map of input to output. This map is the ML model represented as a mathematical function.

The mapping can be straightforward or convoluted and is also called a neural network. For example, clicking a button to switch on a light bulb is an easy problem that a simple neural network can accomplish. The input is the click, and the output is the state of the light bulb (on/off). Learning this sort of mapping is more straightforward and sometimes referred to as shallow learning. For more complex cases like predicting the price of a house based on input parameters like size, the number of rooms, location, distance from schools, etc., learning happens in multiple steps. Each step is a layer in the neural network. Many of these steps are not known and are called hidden layers. Learning with multiple layers is known as deep learning.

---

[4] Russell, Stuart. *Human Compatible*. Allen Lane, an imprint of Penguin Books, 2019.

[5] Ford, Martin R. *Architects of Intelligence*. Packt, 2018.

In ML, depth refers to the number of layers in the neural network. Networks with more than three layers are generally considered deep. While the term *neural networks* might sound brainlike, neural nets are not trying to imitate the brain, but they are inspired by some of its computational characteristics, at least at an abstract level.[6]

DL is uniquely suited to build AIs that often appear human-like or creative, like restoring black-and-white images, creating art, writing poetry, playing video games, or driving a car. Deep learning and neural networks is a broad field and the foundation of many popular AI techniques.

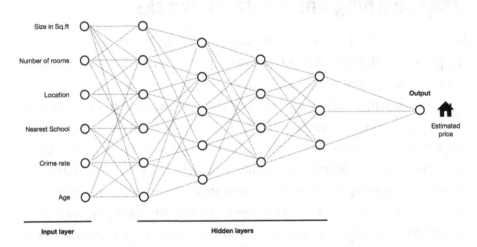

Deep learning and Neural Networks

*Figure 9-7. **Deep learning and neural networks.** In this example, features of a house like its size, number of rooms, location, etc. are used to estimate its price*

---

[6] Ford, Martin R. *Architects of Intelligence*. Packt, 2018.

Backpropagation

A large part of what machine learning teams do is called "feature engineering." Learning is enabled by a collection of what are called "hyperparameters"—an umbrella term that refers to all the aspects of the network that need to be set up by humans to allow learning to even begin.[7] This can include how much weight we should assign for different inputs, the number of layers in a neural network, and many other variables. This is called tuning the algorithm.

An ML engineer might build a model and assign some weights to inputs and different layers. After testing the model on some data, they might adjust these weights to get better results, and they might do this multiple times to arrive at an optimal weight.

By using the backpropagation algorithm, ML teams can start adjusting weights automatically. It's a way of tinkering with the weights so that the network does what you want. As a neural network is trained, information propagates back through the layers of neurons that make up the network and causes a recalibration of the settings (or weights) for the individual neurons. The result is that the entire network gradually homes in on the correct answer.[8]

Backpropagation is used wherever neural networks are employed, especially for speech recognition and speech synthesis, handwriting recognition, and face recognition systems.

---

[7] Mitchell, Melanie. *Artificial Intelligence.* First ed., Farrar, Straus and Giroux, 2019.
[8] Ford, Martin R. *Architects of Intelligence.* Packt, 2018.

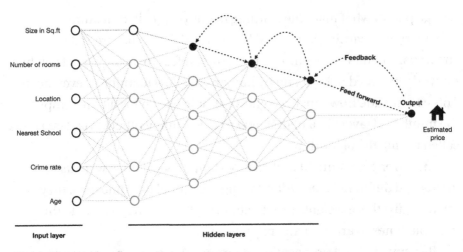

Figure 9-8. Backpropagation. *In this example, features of a house like its size, number of rooms, location, etc. are used to estimate its price. Information or feedback propagates back through the layers of neurons that make up the network and causes a recalibration of the model's weights*

Transfer Learning

If a person knows how to drive a hatchback car, they can also drive an SUV. Software engineers who know a particular language like JavaScript can transfer some of their knowledge when writing code in Python. If I give you a new task, you wouldn't be completely terrible at it out of the box because you'd bring some knowledge from your past experiences with handling similar tasks.

Transfer learning is the ability of an AI program to transfer what it has learned about one task to help it perform a different, related task.[9] Humans are great at doing this. For people, transfer learning is automatic. In

---

[9] Mitchell, Melanie. *Artificial Intelligence.* First ed., Farrar, Straus and Giroux, 2019.

transfer learning, instead of training your model from scratch, you can pick up some learnings from another trained model that is similar. For example, an image recognition system trained to recognize cars from images can also transfer its learnings to recognize golf carts. This technique is very valuable, and many computer vision and natural language processing (NLP) systems are built using transfer learning.

Transfer learning

Figure 9-9. Transfer learning. An image recognition system trained to recognize cars from images can transfer its learnings to identify buses. (a) Image source: https://pythonrepo.com/repo/ MaryamBoneh-Vehicle-Detection. (b) Image source: https:// medium.com/analytics-vidhya/object-detection-zoo-part-1- bus-detection-heavy-vehicle-detection-1f23a13b3c3

Generative Adversarial Networks (GANs)

Perhaps you've seen examples of AI capable of generating images of people who don't really exist, AI-generated speech that sounds human-like, or AI restoring black-and-white pictures to color. These are all examples of generative adversarial networks or GANs in action. Researchers at Nvidia used GANs to generate photorealistic images of fake celebrities from pictures of famous people on the Internet.

Generative adversarial networks were developed by Ian Goodfellow and have been called the most interesting idea of the decade by Facebook's chief AI scientist, Yann LeCun.[10] While explaining how GANs work is out of the scope for this book, at a high level, this technique is a subset of deep learning that generates new data by pitting two AI algorithms against each other, hence adversarial.

Deepfakes, where a person in a specific kind of media—like an image, video, sound, etc.—is swapped with another person, are one of the most popular (and dangerous) applications of GANs. GANs have many applications in the VFX and entertainment industries. You can use GANs for various tasks like synthesizing new images from scratch, generating videos (or deepfakes), restoring images, self-driving cars, video game AIs, and sophisticated robotics.

Generative Adversarial Networks (GANs)

(a) AI generated celebrity images

(b) Painting style transfer on an image

Figure 9-10. *GANs.* (a) *Researchers from Nvidia used artificial intelligence to generate high-res fake celebrities. Source:* www.theverge.com/2017/10/30/16569402/ai-generate-fake-faces-celebs-nvidia-gan. (b) *Style transfer GANs can apply the style from one painting to the other image. Source:* https://softologyblog.wordpress.com/2019/03/31/style-transfer-gans-generative-adversarial-networks/

---

[10] Wiggers, Kyle. "Generative Adversarial Networks: What Gans Are and How They've Evolved." VentureBeat, 2019, https://venturebeat.com/2019/12/26/gan-generative-adversarial-network-explainer-ai-machine-learning/.

Knowledge Graphs

When you search something like "Mumbai" on Google Search, you'll notice that Google seems to understand that Mumbai is a place and will show you a rich view of places to visit, flights, hotels, etc. This rich view is enabled by a technique called knowledge graphs.

A knowledge graph, also known as a semantic network, represents a network of real-world entities—that is, objects, events, situations, or concepts—and illustrates the relationship between them.[11] These rich graphs of information are foundational to enabling computers to develop an understanding of relevant relationships and interactions between people, entities, and events.[12] Knowledge graphs provide a lot of economic value and are probably one of the most underrated AI techniques.

---

[11] "What Is a Knowledge Graph?" ibm.com, 2021, www.ibm.com/cloud/learn/knowledge-graph.

[12] Smith, Brad, and Harry Shum. *The Future Computed*. Microsoft Corporation, 2018.

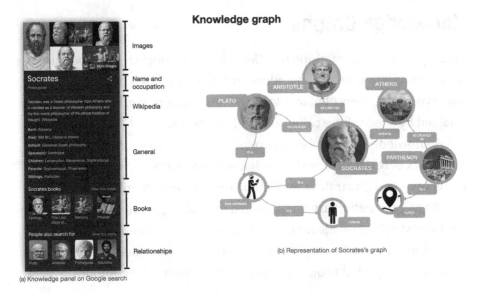

(a) Knowledge panel on Google search

(b) Representation of Socrates's graph

Figure 9-11. Knowledge graphs. *(a) Knowledge panel on Google Search. Source: Google Search on desktop. (b) Representation of Socrates's knowledge graph. Source: https://towardsdatascience. com/knowledge-graphs-at-a-glance-c9119130a9f0*

Note   The preceding list contains mainstream AI techniques and is not exhaustive. Some of the techniques described might be subsets of others. AI is an evolving field, and the line between many techniques is blurry. Sometimes researchers might combine two or more approaches to invent a new technique. For example, the DeepMind group combined reinforcement learning—particularly Q-learning—with deep neural networks to create a system that could learn to play Atari video games. The group called their approach deep Q-learning.[13]

---

[13] Mitchell, Melanie. *Artificial Intelligence.* First ed., Farrar, Straus and Giroux, 2019.

AI Metrics

AI systems are probabilistic and will sometimes make mistakes. Lack of complete information is an insurmountable challenge with real-world AI systems. Very little of our knowledge is completely certain; we don't know much about the future. However, complete certainty is not necessary for action. We only need to know the best possible action given the circumstances. In other words, it is much more important for AI models to be useful than perfect. AI product teams use some common metrics and terms to measure the usefulness of their models.

Complete certainty is not necessary for action.

To determine the success and failure of a model's outputs, we need to know when it succeeds or makes mistakes and what types of mistakes it makes. To understand this, let's take an example of a system that detects donuts and "no donuts." We can think of detecting a donut as a positive result and detecting "no donut" as negative. There are four possible outcomes:

1. **True positive**

   The model correctly predicts a positive result, that is, the system detects a donut correctly, and the image actually contains donuts.

2. **True negative**

   The model correctly predicts a negative result, that is, the AI detects "no donut" correctly, and the image doesn't contain a donut.

3. **False positive**

   The model incorrectly predicts a positive result, that is, the system detects a donut when there is no donut in the image.

4. **False negative**

The model incorrectly predicts a negative result,
that is, the AI predicts "no donut" when the image
actually contains a donut.

We can plot the preceding outcomes in a table. This table is called the
confusion matrix and is used to describe the performance of a model.
Let's assume we give this detector 75 images of donuts and 25 photos with
"no donuts" in them. In our early tests, we find out that of the 75 images of
donuts, our detector recognizes 70 correctly (it's donut) and 5 incorrectly
(it's "not donut," but the detector says it's donut). When classifying the 25
images of "no donut," our detector recognizes 21 correctly (it's "no donut")
and 4 incorrectly (it's a donut, but the detector says it's "no donut").

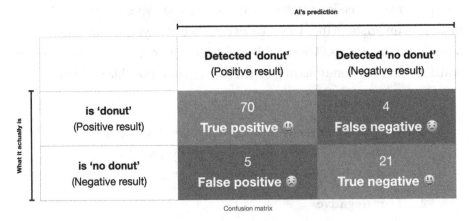

Figure 9-12. Confusion matrix of the "donut" detector

Once we know these values, we can use them to derive some standard
metrics used by AI teams to understand the performance of their models.

Accuracy

Accuracy is the fraction of predictions our model got right. It is the proportion of correct predictions out of all predictions.

Accuracy = (True positives + True negatives) ÷ All predictions*

*All predictions = True positives + True negatives + False positives + False negatives

In the case of our donut detector, that would be

Accuracy = (70 + 21) / (70 + 5 + 4 + 21) = 0.91 (or 91%)

91%! That's not so bad. However, there are problems with determining an algorithm's performance based on accuracy alone. A 91% accuracy may be good or terrible depending on the situation. (It's not good if a car crashes 9 out of 100 times).[14] To understand the AI's performance better, AI teams use metrics like precision and recall that describe the breadth and depth of results that your AI provides to users and the types of errors that users see.[15]

Precision

Precision is the proportion of true positives correctly predicted out of all true and false positives. Precision determines how much you should trust your model when it says it's found something. The higher the precision, the more confident you can be that any model output is correct.[16] However, the tradeoff with high precision is that the model might miss out on predicting some correct results, that is, you will increase the number of false negatives by excluding possibly relevant results.

---

[14] Kore, Akshay. "Designing AI products." woktossrobot.com, 2020, https://woktossrobot.com/aiguide/.

[15] pair.withgoogle.com, https://pair.withgoogle.com/.

[16] pair.withgoogle.com, https://pair.withgoogle.com/.

Precision = True positives ÷ (True positives + False positives)

In the case of our donut detector, that would be

Precision = 70 / (70 + 5) = 0.933 (or 93.3%)

If the donut detector was optimized for high precision, it wouldn't recommend every single image of a donut, but it would be highly confident of every donut it recommends. Users of this system would see fewer incorrect results but would miss out on some correct predictions.

Recall

Recall is the proportion of true positives correctly predicted out of all true positives and false negatives. The higher the recall, the more confident you can be that all the relevant results are included somewhere in the output.[17] The higher the recall, the higher is the overall number of predictions. However, the tradeoff with high recall is that the model might predict more incorrect results, that is, you will increase the number of false positives by including possibly irrelevant results.

Recall = True positives ÷ (True positives + False negatives)

In the case of our donut detector, that would be

Recall = 70 / (70 + 4) = 0.945 (or 94.5%)

If we optimized the donut detector for high recall, it would recommend more images of donuts, including ones that don't contain donuts. The user of this system would see more results overall; however, more of those results might be incorrect.

---

[17] pair.withgoogle.com, https://pair.withgoogle.com/.

Precision vs. Recall Tradeoff

In most real-world scenarios, you will not get a system that both is completely precise and has a hundred percent recall. Your team will need to make a conscious tradeoff between the precision and recall of the AI. You will need to decide if it is more important to include all results even if some are incorrect, that is, high recall, or minimize the number of wrong answers at the cost of missing out on some right ones, that is, high precision. Making this decision will depend on the context of your product and the stakes of the situation. Weighing the cost of false positives and false negatives is a critical decision that will shape your users' experiences.[18] For example, in a banking system that classifies customers as loan defaulters or not, you might be better off making fewer but highly confident predictions. It would be better to optimize this system for higher precision. On the other hand, a music streaming app occasionally recommending a song that the user doesn't like isn't as important as showing a large selection. You might optimize such a system for higher recall.

You will need to design your product's experiences for these tradeoffs. Make sure to test the balance between precision and recall with your users.

---

[18] pair.withgoogle.com, https://pair.withgoogle.com/.

Precision vs. Recall

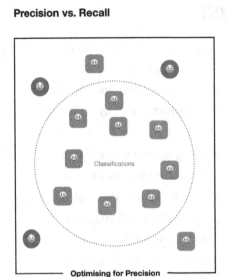
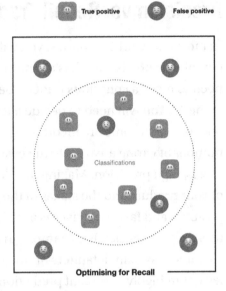

Figure 9-13. Precision vs. recall tradeoff. (Left) Optimizing for precision: Model classifies no false positives but misses some true positives. (Right) Optimizing for recall: Model covers all true positives but includes some false positives

---

Note   The preceding are some of the most popular metrics AI teams use. However, there can be various additional metrics specific to certain AI techniques.

AI Capabilities

The AI techniques discussed previously enable ML teams to build capabilities that help product teams solve various user problems. The following are some of the most common capabilities used to build AI products:

1. Computer vision (CV)

2. Natural language processing (NLP)

3. Speech and audio processing

4. Perception, motion planning, and control

5. Prediction

6. Ranking

7. Classification and categorization

8. Knowledge representation

9. Recommendation

10. Pattern recognition

Computer Vision (CV)

Computer vision (CV) is the ability of AI systems to see by enabling them to derive meaningful information from digital images, videos, and other visual inputs. AI systems can use this information to make predictions and decisions and take action. For example, Airware was a start-up that conducted an automated analysis of drone images for mining and construction industries. They used computer vision on drone imagery to analyze safety compliance, maximize fuel efficiency, calculate stockpile volumes, and recognize storm damage.[19] Google Lens uses computer vision algorithms to identify what's in a picture.

---

[19] a16z. AI: What's Working, What's Not. 2017, www.youtube.com/watch?v=od7quAx9nMw.

The following are some key applications of computer vision:

1. **Image classification**

   This is the ability to classify images by parameters like what the image contains, objects, people, and other information. For example, an intelligent photo album might use image classification to categorize images into landscapes, photos of people, animals, etc.

2. **Face recognition**

   Face recognition is a method of identifying an individual using their face. While these systems raise ethical concerns, facial recognition is fairly common, from iPhone's Face ID to government security and surveillance.

3. **Object detection**

   This is the ability to detect objects in an image, for example, detecting the position of cars and pedestrians in an image. They can be used to tell where objects are present in an image by drawing a polygon, often a rectangle, around them. Object detection is used in many applications like self-driving cars, radiology, counting people in crowds, robotic vacuum cleaners, and even precision farming.

4. **Image segmentation**

   Image segmentation is similar to object detection, but you can segment an exact boundary instead of drawing a rectangle around the detected object in an image or video. This technique is commonly used in VFX as well as many photo and video editing applications. The ability to blur your background on a video call or change the background to a beach is enabled by image segmentation.

5. **Object tracking**

This is the ability to track the movement of objects in a video. You can detect where things are going, like people on the camera feed, microorganisms in a cell culture, traffic, etc.

Figure 9-14. Computer vision applications: *image classification, object detection, and image segmentation. Source:* https:// bdtechtalks.com/2021/05/07/attendseg-deep-learning-edge-semantic-segmentation/

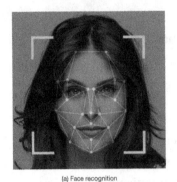 

(a) Face recognition   (b) Object tracking

Figure 9-15.   Computer vision applications. (a) Face recognition. Source: https://blogs.microsoft.com/on-the-issues/2020/03/31/washington-facial-recognition-legislation/. (b) Object tracking of vehicles in the camera footage. Source: https://aidetic.in/blog/2020/10/05/object-tracking-in-videos-introduction-and-common-techniques/

Natural Language Processing (NLP)

Natural language processing (NLP) is the ability for AI to pull insights and patterns out of written text. NLP is one of the most valuable capabilities for AI systems and is used in many applications like AI assistants, search engines, chatbots, sentiment analysis, translation, etc. For example, voice assistants like Alexa, Siri, Cortana, or Google Assistant use NLP to act on your commands. Everlaw is a cloud-based AI application that helps lawyers research documents and prepare for trials by extracting information from legal documents, highlighting which documents or their parts are important. They automate repetitive tasks for lawyers and paralegals by identifying documents relevant for trial by topic, translating documents into other languages, clustering documents by topics, and audio transcription.[20]

---

[20] a16z. AI: What's Working, What's Not. 2017, www.youtube.com/watch?v=od7quAx9nMw.

The following are some mainstream applications of NLP:

1. **Text classification**

   This is the ability to categorize text by its contents. For example, a spam detection tool would use the email's subject and content to classify it as spam or not spam. Text classification can also be used to automatically categorize products by their description or translate restaurant reviews to star ratings.

2. **Information retrieval**

   Information retrieval enables systems to extract information from unstructured text in documents or websites. This technique is used in various search applications like web search, finding documents on your computer, searching for people on social media, etc.

3. **Entity recognition in text**

   This is the ability to recognize entities in text. For example, you can use entity recognition to find and extract names, places, addresses, phone numbers, dates, etc. from textual information. Entity recognition is commonly used in chatbots to extract information like names, addresses, and queries to decide appropriate responses.

4. **Machine translation**

   Machine translation is used to translate text from one language to another. For example, Facebook uses machine translation to translate comments on posts. Google Translate uses this technique to translate between languages. The idea of machine translation is sometimes extended to different input-output mappings like translating words to images, sounds, and other forms of media.

5. You can also use NLP to tag parts of speech like nouns, verbs, etc., sentiment analysis, etc. Grammarly is a writing assistant that uses various NLP techniques to identify mistakes and which parts of the text to change, highlight which parts to pay attention to, etc.

(a) Text classification

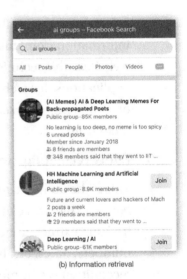

(b) Information retrieval

Figure 9-16. NLP applications. *(a) Text classification: The content of the review is automatically classified as a rating. (b) Information retrieval: Searching for groups on Facebook. Source: Facebook app*

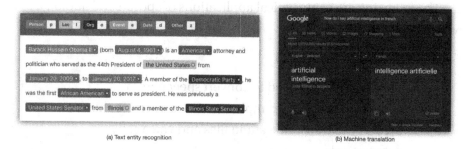

(a) Text entity recognition   (b) Machine translation

Figure 9-17. NLP applications. (a) Text entity recognition: The algorithm identifies and tags entities like people, locations, events, etc. in the paragraph. Source: www.analyticsvidhya.com/blog/2021/11/a-beginners-introduction-to-ner-named-entity-recognition/. *(b) Machine translation. Source: Google Translate*

Speech and Audio Processing

This is the ability for AI to convert speech to text and text to speech (TTS) and extract information from audio files. Speech processing is often used alongside NLP in various AI applications like voice assistants, call center analytics, automatic subtitle generation, etc. For example, voice assistants like Alexa, Siri, Cortana, or Google Assistant use speech processing to recognize and convert your voice into text. YouTube uses speech processing to generate automatic subtitles on videos. Spleeter is an open source software made by the music streaming service Deezer that uses speech and audio processing to automatically separate vocals and different instruments from songs.

The following are some typical applications of speech and audio processing:

1. **Speech to text (STT)**

   This technique is used to convert speech from audio into text. For example, Otter.ai is a service that uses STT to transcribe audio files and voice calls automatically.

2. **Wake word detection**

   A wake word is a special word or phrase meant to activate a device when spoken.[21] It is also known as a trigger word or hot word. Voice assistants like Siri or Amazon's Alexa use wake word detection to identify phrases like "Hey, Siri" or "Alexa" to start interacting with them.

3. **Speaker ID**

   This is the ability of AI systems to identify who is speaking. Speaker ID is commonly used in automatic transcription tools, voice assistant devices meant for multiple users, and voice authentication.

4. **Audio entity recognition**

   Audio entity recognition detects various types of entities in audio files like different instruments, people's voices, animal sounds, etc. This technique is commonly used in music production to separate instruments and vocals. Some advanced home

---

[21] "Tips for Choosing a Wake Word." picovoice.ai, https://picovoice.ai/docs/tips/choosing-a-wake-word/.

theater systems also use some form of audio entity detection to fire different types of sounds from different speakers and simulate surround sound. Audio detection can also be used for ensuring safety in industrial environments by detecting a gas leak, pipe bursts, etc.

5. **Speech synthesis**

Also known as text to speech (TTS), this is the ability to generate speech from text. This capability is commonly used in voice assistants to respond to user commands or provide information. It is also valuable for accessibility applications like read-aloud. TTS eases the Internet experience for one out of five people who have dyslexia, low-literacy readers, and others with learning disabilities by removing the stress of reading and presenting information in an optimal format.[22]

---

[22] Lacomblez, Courtney. "Who Uses Text to Speech (TTS) Anyway?" ReadSpeaker, 2017, www.readspeaker.com/blog/uses-text-speech-tts-anyway/.

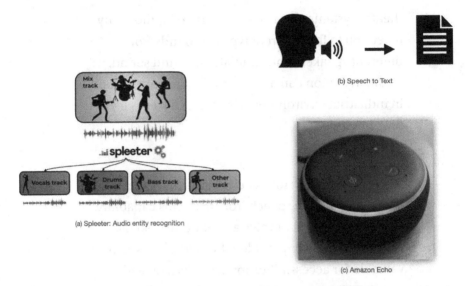

Figure 9-18. *Speech and audio processing applications. (a) Audio entity recognition: Spleeter's service automatically splits an audio track by vocals, instruments, and other sounds. Source:* `https://deezer.io/releasing-spleeter-deezer-r-d-source-separation-engine-2b88985e797e`. *(b) Speech to text. (c) Alexa on Amazon Echo uses multiple techniques like wake word detection, speech to text, speech synthesis, and speaker ID, along with other AI capabilities. Source: Photo by Lazar Gugleta on Unsplash*

Perception, Motion Planning, and Control

Robotics is a branch of technology that deals with the design, construction, operation, and application of robots. Primarily used in robotics, this is the ability of AI systems to plan and navigate spaces through sensors and control actuators. An actuator is a component of any machine that enables movement like a robot's limbs, steering wheel in a self-driving car, propellers on a drone, etc. Perception helps a robot figure out what's around it, motion planning allows the robot to map the best route, and control allows it to send commands to actuators. Robots are used in many industrial applications like warehouse robots, assembly lines in factories, and delivering products using drones.

While the most common illustration of a robot is one that resembles the human form, our homes, factories, and surroundings are filled with different types of robots. Here are some examples:

1. A robotic vacuum cleaner would scan its environment to create a map of a room along with obstacles. It would then plan its motion to navigate and clean the space in the most optimal manner. It would use its vacuum attachments to perform the task of cleaning the room.

2. Zipline is a start-up that uses self-flying drones to drop off blood where it is needed. This is especially useful in emergency scenarios where an ambulance or truck might take a long time to reach or it is dangerous to get to the place.[23]

3. Amazon has more than 200,000 mobile robots working inside its warehouse network.[24] Many of these robots carry shelves of products from worker to worker, read barcodes, and pick up and drop off items from and on the conveyor belt.

4. Self-driving vehicles are essentially autonomous robots on wheels. They use various sensors like

---

[23] a16z. AI: What's Working, What's Not. 2017, www.youtube.com/watch?v=od7quAx9nMw.

[24] Del Rey, Jason. "How Robots Are Transforming Amazon Warehouse Jobs—For Better and Worse." vox.com, 2019, www.vox.com/recode/2019/12/11/20982652/robots-amazon-warehouse-jobs-automation.

cameras, lidar scanners, etc. to build a map of
their surroundings, do motion planning, and use
actuators like the steering wheel, headlights, brakes,
etc. to navigate.

(a) Robotic vacuum cleaner   (b) Robots in a factory   (c) Google's self driving car

Figure 9-19.   Examples of robots using perception, motion planning, and control. *(a) Robotic vacuum cleaner. Source: Photo by YoonJae Baik on Unsplash. (b) Robots in a factory. Source: Photo by David Levêque on Unsplash. (c) Google's self-driving car. Source:* `https://commons.wikimedia.org/wiki/File:Google_self_ driving_car_at_the_Googleplex.jpg`

Prediction

Prediction is the process of filling in the missing information. Prediction takes the information you have, often called "data," and uses it to generate information you don't have.[25] This is one of the most valuable and powerful capabilities of AI systems. For example, AI models are used in systems that predict flight prices or the number of orders while managing inventory. A ubiquitous example is the smartphone keyboard that uses predictive models to suggest the next word. Furthermore, predictions can

---

[25] Agrawal, Ajay, et al. *Prediction Machines*. Harvard Business Review Press, 2018.

be used to make an investment, detect and correct health disorders, or choose the right word in a situation.

The following are some examples of predictions:

1. Airbnb predicts the probability of getting a booking based on price, date, photos, location, and unique listing features.

2. Cardiogram is a personal healthcare assistant that detects atrial fibrillation and predicts strokes using data from a smartwatch.[26]

3. Instacart's partner app helps personal shoppers predict the best route in-store to collect groceries as fast as possible.

Ranking

You can use AI to rank items, especially when it is difficult to determine a clear ranking logic. Ranking algorithms are used in search engines to decide the order of results. PageRank is an algorithm used by Google Search to rank web pages in search results. Ranking is also used along with other AI applications like product recommendation systems to decide the order of items suggested.

---

[26] a16z. AI: What's Working, What's Not. 2017, www.youtube.com/watch?v=od7quAx9nMw.

Classification and Categorization

This is the ability of AI to categorize entities into different sets. Categorization can be used for several applications like sorting vegetables from a pile or detecting faulty products in an assembly line or by photo apps to classify images into landscapes, selfies, etc. Clustering is a common technique used for generating a set of categories. For example, in an advertising company, you can use clustering to segment customers based on demographics, preferences, and buying behavior.

Knowledge Representation

Knowledge representation is the ability of AI systems to extract and organize information from structured and unstructured data like web pages, books, databases, real-world environments, etc. This capability enables AI systems to understand relevant relationships and interactions between people, entities, and events. A rich search engine results page (SERP) is an example of knowledge representation in action. For example, when you search for a famous person like "Alan Turing" on Google Search, the result shows you rich information about his date of birth, partner, education, alma mater, etc.

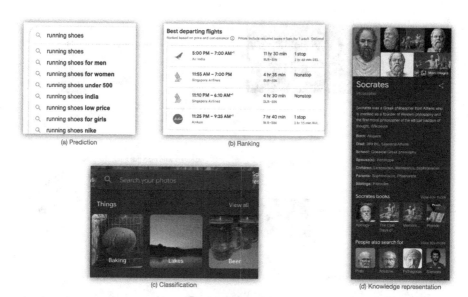

Figure 9-20. *(a)* **Prediction:** *Google Search automatically predicts the query. Source: Google search on Desktop. (b)* **Ranking:** *Flight booking website ranks results by relevance. (c)* **Classification:** *Google Photos automatically sorts images by things in them. Source: Google Photos app on iOS. (d)* **Knowledge representation:** *Knowledge panel on Google Search. Source: Google Search on Desktop*

Recommendation

Recommendation is the ability of AI systems to suggest different content to different users. For example, Spotify suggests what songs to listen to next, or Amazon recommends which books to buy based on previous purchases and similar buying patterns. Recommendation is a very valuable capability and is closely related to prediction. You can also use recommendations to surface content that would otherwise be impossible for users to find on their own. Many social media applications like TikTok or Instagram use recommendation algorithms to generate a continuous stream of dynamic content.

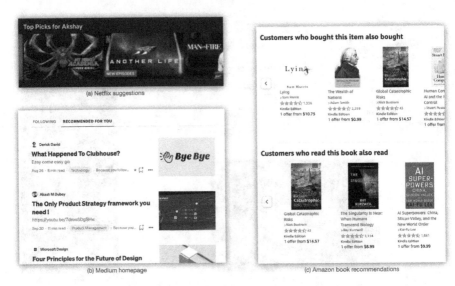

Figure 9-21. Recommendation. *(a) Netflix allows users to choose a title from their personalized suggestions. Source:* $www.netflix.com/$*. (b) Users can choose from articles recommended for them on Medium. Source:* $https://medium.com/$*. (c) Amazon shows multiple choices of results in its book suggestions. Source:* $https://amazon.in/$

Pattern Recognition

This is the ability of AI systems to detect patterns and anomalies in large amounts of data. Detecting anomalies means determining specific inputs that are out of the ordinary. For example, AI systems can help radiologists detect lung cancer by looking at X-ray scans. Banks use software that detects anomalous spending patterns to detect credit card fraud. Market researchers can also use pattern recognition to discover new segments and target users.

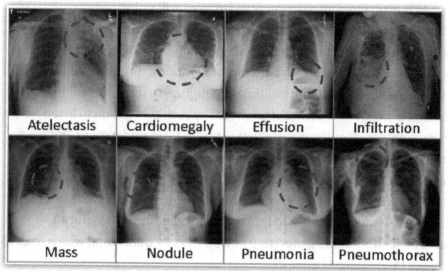

Deep learning based X-ray image analysis

***Figure 9-22.   Pattern recognition.** Deep learning–based X-ray image analysis. Source:* `www.semanticscholar.org/paper/Deep-learning-based-medical-X-ray-image-recognition-Khan/fdd644ec07a4e54d1860825f07f74da2f6f8e442`

Apart from the preceding capabilities, machine learning is also used broadly to find patterns in data. Many of the examples discussed previously relate to unstructured data. The popular press frequently covers AI advancements applied on unstructured data like detecting cats from videos, generating music, writing movie scripts, etc. These examples often appear human-like and are easily understandable by most people.

However, AI is applied at least as much or more to structured data, that is, data is a standardized format for storing and providing information. Structured data also tends to be more specific to a single organization. AI on structured data is creating immense economic value to organizations. But since it is harder for popular media to understand AI advancements on structured data within companies, it is written about much less than AI advancements over unstructured data.

AI is evolving, and many new techniques will unlock newer capabilities. A large part of what AI teams do is figure out problems to solve with AI and match appropriate techniques and capabilities to the solution. Doing this well requires lots of iterations, time, and effort. Most large AI products combine multiple capabilities to provide a solution. For example, an AI voice assistant might use speech processing to listen for voice commands and convert them to text, NLP to understand and act on the command, and, finally, speech synthesis to respond to the user. Many of the preceding capabilities might be further combined to realize new AI methods. For example, you could extend the idea of machine translation from translating languages to translating images into words through automatic captioning, generating poetry, or even translating words back to images by creating AI artworks.

Summary

Product designers and product managers need to understand what their tech teams do and speak their language to work effectively within AI teams. In this chapter, we looked at some mainstream AI techniques, metrics used to evaluate the success of AI algorithms, and common capabilities of AI. Here are some important points:

1. There are different ways to implement AI, but at a high level, they are based on two approaches, namely, rule-based and examples-based.

    a. In the rule-based approach, the AI is programmed to follow a set of specific instructions.

    b. In the examples-based approach, learning happens by giving the AI a bunch of examples of what it will encounter in the real world and how to respond.

2. A significant amount of progress in AI techniques
    has been made in the subfield of machine
    learning (ML).

3. The following are some of the mainstream AI
    techniques:

    a. **Supervised learning**

        In supervised learning, you train the AI by
        mapping inputs to outputs. We use labeled data
        to build the model.

    b. **Unsupervised learning**

        In unsupervised learning, there is no labeled
        data. We teach machines to learn directly
        from unstructured data coming from their
        environments.

    c. **Reinforcement learning**

        In reinforcement learning, the AI learns
        by rewarding it for positive behavior and
        punishing for negative behavior. The AI learns
        by trial and error.

    d. **Deep learning and neural networks**

        For complex cases like predicting the price of
        a house based on input parameters like size,
        the number of rooms, location, distance from
        schools, etc., learning happens in multiple
        steps. Each step is a layer in the neural network.
        Learning with multiple layers is known as deep
        learning.

e.   **Backpropagation**

In backpropagation, feedback on a neural
network propagates back through the layers of
neurons. This information is used to recalibrate
the model's weights.

f.   **Transfer learning**

Transfer learning is the ability of an AI program
to transfer what it has learned about one task to
help it perform a different, related task.[27]

g.   **Generative adversarial networks (GANs)**

GANs are a subset of deep learning that
generates new data by pitting two AI algorithms
against each other. GANs have many
applications in the VFX and entertainment
industries. You can use GANs for various tasks
like synthesizing new images from scratch,
generating videos (or deepfakes), restoring
images, self-driving cars, video game AIs, and
sophisticated robotics.

h.   **Knowledge graphs**

A knowledge graph, also known as a semantic
network, represents a network of real-world
entities—that is, objects, events, situations, or
concepts—and illustrates their relationship.

---

[27] Mitchell, Melanie. *Artificial Intelligence.* First ed., Farrar, Straus and Giroux, 2019.

4. Lack of complete information is an insurmountable challenge with real-world AI systems. Very little of our knowledge is entirely certain; we don't know much about the future.

5. We only need to know the best possible action given the circumstances. In other words, it is much more important for AI models to be useful than perfect.

6. AI product teams use some common metrics and terms to measure the usefulness of their models:

   a. **Accuracy**

      Accuracy is the fraction of predictions our model got right. It is the proportion of correct predictions out of all predictions.

   b. **Precision**

      Precision is the proportion of true positives correctly predicted out of all true and false positives. Precision determines how much you should trust your model when it says it's found something.

   c. **Recall**

      Recall is the proportion of true positives correctly predicted out of all true positives and false negatives.

7. In most real-world scenarios, you will not get a system that both is completely precise and has a hundred percent recall. Your team will need to make a conscious tradeoff between the precision and recall of the AI.

8. The AI techniques discussed previously enable ML
    teams to build capabilities that help product teams
    solve various user problems. The following are some
    of the most common capabilities used to build AI
    products:

    a. **Computer vision (CV)**

        Computer vision (CV) is the ability of AI
        systems to see by enabling them to derive
        meaningful information from digital images,
        videos, and other visual inputs.

    b. **Natural language processing (NLP)**

        Natural language processing (NLP) is the
        ability for AI to understand natural language by
        pulling insights and patterns out of written text.

    c. **Speech and audio processing**

        This is the ability for AI to convert speech to text
        and text to speech and extract information from
        audio files.

    d. **Perception, motion planning, and control**

        Primarily used in robotics, this is the ability of
        AI systems to plan and navigate spaces through
        sensors and control actuators. Perception helps
        a robot figure out what's around it, motion
        planning allows the robot to map the best route,
        and control allows it to send commands to
        actuators.

e.   **Prediction**

Prediction is the process of filling in the missing information. Prediction takes the information you have, often called "data," and uses it to generate information you don't have.[28]

f.   **Ranking**

You can use AI to rank items, especially when it is difficult to determine a clear ranking logic. Ranking algorithms are used in search engines to decide the order of results.

g.   **Classification and categorization**

This is the ability of AI to categorize entities into different sets.

h.   **Knowledge representation**

Knowledge representation is the ability of AI systems to extract and organize information from structured and unstructured data like web pages, books, databases, real-world environments, etc.

i.   **Recommendation**

Recommendation is the ability of AI systems to suggest different content to different users.

j.   **Pattern recognition**

This is the ability of AI systems to detect patterns and anomalies in large amounts of data.

---

[28] Agrawal, Ajay, et al. *Prediction Machines*. Harvard Business Review Press, 2018.

9. The popular press frequently covers AI advancements applied to unstructured data like detecting cats from videos, generating music, writing movie scripts, etc. However, AI is used at least as much or more on structured data, which tends to be more specific to a single organization and is creating immense economic value.

10. Most large AI products combine multiple capabilities to provide a solution. A large part of what AI teams do is figure out problems to solve with AI and match appropriate techniques and capabilities to the solution.

CHAPTER 10

Working Effectively with AI Tech Teams

Building great AI products requires teamwork. In this chapter, we will discuss how designers can work effectively with their AI tech teams. We will understand what an AI tech team looks like and the behaviors and strategies you could employ to motivate, empathize with, and collaborate effectively with your machine learning and engineering counterparts.

In my final year of architecture school, I realized what a terrible singer I was. During my undergraduate studies, a few of my classmates decided to form a band. We reached out to everyone we thought could help us—people who sing and play the keyboard, the drums, the guitar, and any other instrument. The band's lead member was my buddy, who asked me to sing. I'd received some minor applause at a party at his place. That got to my head, and I said yes. I thought it would be a good experience to be a part of a band. It was pretty evident to everyone except me that I was terrible at it. And because I was a senior, no one called it out. It is one thing to sing at karaoke or a friend's place, but it is another thing altogether to perform in front of an audience. Creating a musical performance is hard work and involves lots of collaboration, skill, and technical knowledge.

Building AI products is technical but also requires creativity. Popular culture depicts people who work in technology as science and math nerds with low EQ. Characters like Steve Urkel on *Family Matters*, Sheldon Cooper on *The Big Bang Theory*, and Dennis Nedry from *Jurassic Park*

A. Kore, *Designing Human-Centric AI Experiences*,
https://doi.org/10.1007/978-1-4842-8088-1_10

all create a picture of social misfits and man-children, more comfortable with a slide rule than a conversation.[1] The media loves to perpetuate these stereotypes, which can be incredibly misleading. While most people in tech are interested in science and technology, the best people I know who build AI products have eventful lives with interesting hobbies. Building AI products is more similar to making music or producing a film than doing math or science.

> *Building AI products is more similar to making music or producing a film than doing math or science.*

An orchestra needs different musicians to communicate, collaborate, and work together to create a great performance. Similarly, different members of your AI product team need to work together to build a great user experience. If you wonder why "simple" changes to your product take so long to make, then start by understanding the interaction between you and your engineering team.

Think of engineers less as people who churn out code and more like creative problem-solvers. Building a great user experience is everyone's responsibility. Extend your UX family. As a product designer or product manager, getting too prescriptive too quickly may result in unintentional anchoring and diminish the creativity of your engineering counterparts. There are many ways to approach any AI problem, and your engineers love overcoming technical challenges. Trust them to use their intuition and encourage them to experiment.[2]

---

[1] Lawson, Jeff, and Eric Ries. *Ask Your Developer*. Harper Business, 2021.

[2] Holbrook, Jess. "Human-Centered Machine Learning." Medium, 2017, `https://medium.com/google-design/human-centered-machine-learning-a770d10562cd`.

Common Roles in an AI Product Team

When trying to build an AI product, you'll rarely be working in silos. The complexity and the number of decisions required to create great AI products necessitate multiple individuals working in teams. There are different types of work needed to build AI products. So far, job titles aren't well defined, and many functions overlap.

The following are the most common roles in an AI product team.

Machine Learning Engineer

ML engineers build, train, test, and deploy AI models. They also are responsible for gathering the data required to train AI. ML engineers also build APIs for other teams like engineering, data science, etc. to consume for building applications or analysis.

Machine Learning Researcher

AI is an evolving field. ML researchers work toward extending the state of the art in AI. They conduct research by experimentation and invention, going through the academic literature, publishing their research in conferences, and even patenting novel techniques.

Applied ML Scientist

You can think of an ML scientist as someone who does the work of an ML engineer and ML researcher. They go through the academic literature and research papers to find state-of-the-art techniques. ML scientists take these techniques and translate them into AI models.

Software Engineer

Software engineers write software to build the AI product, connect the AI model through APIs, and conduct security, reliability, and crash tests. Software engineers are most often divided into frontend and backend teams. Frontend teams work on building the user interface, while backend teams work on the underlying infrastructure.

This is a common role in all AI teams, and sometimes more than half your team might be software engineers. Depending on the maturity of your engineering department, software engineering might be further broken down into teams that handle different parts of the software development process like front end, back end, security, quality assurance (QA), infrastructure, etc.

Data Engineer

AI products require and generate large amounts of data. Sometimes data volumes can be so large that it requires a lot of work to manage them. For example, a self-driving car might store 1 GB of data every minute. It is not uncommon for large companies to generate multiple petabytes of information every day.

Data engineers help organize data and make sure that it is saved in an easily accessible, secure, and cost-effective way. Handling and managing data is a significant cost for AI products, and data engineers can help you significantly reduce infrastructure costs. Sometimes, ML engineers also do the job of data engineering.

Data Scientist

The role of a data scientist is still evolving and not well defined. Different organizations might assign different responsibilities to their data science teams. However, in most cases, data scientists examine data and provide insights to various teams like product management, marketing, sales, engineering, etc. They will often interface with your leadership team by making presentations around their insights and recommendations. In some organizations, data scientists also do the job of ML engineers.

Product Manager

Product managers help decide what to build and what's feasible, desirable, and valuable. They help prioritize features and products and identify, test, and validate product-market fit. They collaborate with different stakeholder teams in the organization like engineering, design, sales, marketing, finance, legal, and leadership.

Sometimes, your organization might have a separate role of an AI product manager or a technical product manager (TPM). An AI product manager is someone who understands the technical aspects of AI as well as business needs.

Product Designer

If you're reading this book, you're probably familiar with this role. Product designers are responsible for the user experience of your AI product, usually taking direction from product management, business goals, leadership, and overall company objectives. The product design discipline is broad and varies across companies. In mature organizations, product design teams may be broken down into user experience design, visual design, UI design, usability, user research, etc.

Your job as a designer is to help your product team make great user-centered choices.

Effective Collaboration

At its heart, building AI products is a creative endeavor. You are trying to solve a user problem by building not just the software but also imbuing it with intelligence. Training an AI model is a slow and iterative process. Engineers often need to use their judgment and imagination when tuning the algorithm. Your job as a designer is to help your product team make great user-centered choices along the way.[3]

Collaboration is a two-way activity. Inspire your team members with stories, prototypes, customer feedback, and findings from user research and introduce them to UX methods and design principles. Sharing the design process can help your team be comfortable with iteration and work wonders for your ability to influence the product. Mostly, you just have to treat your team members as people, full of ambitions to learn and grow, motivations to do their best work, and a range of skills they want to exercise.

> *Our attempts to create artificial intelligence have, at the very least, helped highlight the complexity and subtlety of our own minds.*

Easy Things Are Hard

Many of the things that children can easily do, like picking up objects of different shapes, recognizing the tone of voice, navigating playgrounds, etc., have turned out to be surprisingly difficult for AI systems to achieve than seemingly complex tasks like diagnosing diseases, beating human champions at chess and *StarCraft*, or solving complex algebraic problems. Easy things are hard. There's a famous rule of thumb in any complex engineering project: the first 90% of the project takes 10% of the time, and

---

[3] Holbrook, Jess. "Human-Centered Machine Learning." Medium, 2017, https://medium.com/google-design/human-centered-machine-learning-a770d10562cd.

the last 10% takes 90% of the time.[4] And while child's play is hard, our most critical jobs such as running organizations and governments, caregiving, or teaching chemistry that have mostly complex and unobservable environments are even harder for AI. Our attempts to create artificial intelligence have, at the very least, helped highlight the complexity and subtlety of our own minds.

Collaborate; Don't Dictate

Most designers I know cringe at the idea of someone sitting next to them and asking them to change fonts, colors, and other elements on a design artifact. To me, nothing is more annoying than someone breathing down your throat and micromanaging your work by telling you exactly what to do. It might be OK to do it in some cases, but most times, I'd like to be left alone to do focused work. Apart from demotivating people, micromanagement signals that you don't trust them to make decisions.

Many of us do the same when it comes to working with tech teams. After building a proposal, getting buy-in from leadership, we go to our engineers and say something like "Hey, do this quick, by this deadline" and then run away. Engineering teams then come back with a timeline that you did not account for. Seemingly "simple" changes take much longer than anticipated. Your tech team does not seem as excited, engaged, or motivated. The relationship between developers and businesspeople is not well understood, but is critical to solving business problems with technology.[5]

Apart from demotivating people, micromanagement signals that you don't trust them to make decisions.

---

[4] Mitchell, Melanie. *Artificial Intelligence*. First ed., Farrar, Straus and Giroux, 2019.
[5] Lawson, Jeff, and Eric Ries. *Ask Your Developer*. Harper Business, 2021.

Saying things like "This should just be a quick thing" or "You can do this in a day" is not only annoying but also meaningless. Unless you understand how things are put together, unless you know how the infrastructure is built, you have no idea how long it will take.[6] How can product teams commit to a deadline if they don't understand how long something will take to build? And not having enough time to build will lead to one or more of these undesirable outcomes:

- Features will get cut to meet the deadline.

- Product quality will suffer.

- Your team will be burnt out.

Share Problems, Not Solutions

The key to getting product teams and developers to work well together is for the product teams to share problems, not solutions. Instead of presenting engineers with a solution that's already defined, product designers and managers can share the problem and ask engineers to help figure out the fastest way to solve it based on how the existing systems are constructed. Including your tech teams early in the discussion is not just about being nice or not hurting their feelings; it also has serious advantages. Many developers have some deeper insight into the integration or feasibility of some products or some features. Finding the shortest technical path in context is what engineers do for a living. It's what they're trained to do in computer science classes.[7] But instead of harnessing their knowledge and insight, most companies just tell engineers precisely what to do. They ask them to take off their creative and problem-solving hats. Don't tell; ask. Share problems, not solutions.

Don't tell; ask. Share problems, not solutions.

---

[6] Lawson, Jeff, and Eric Ries. *Ask Your Developer*. Harper Business, 2021.

[7] Lawson, Jeff, and Eric Ries. *Ask Your Developer*. Harper Business, 2021.

The key to building great AI products is to engage your product and tech teams in leveraging their whole brain toward the problem. Design is creative, and so is code. "I've always thought that engineering is one of the most creative jobs in the world," Amazon's CTO, Werner Vogels, says.[8] Yet many teams don't realize that writing code is creative, they don't create an environment where developers can exercise this creative muscle, and everybody loses. Trust that your tech teams are adept at applying the craft of software to serving customers and solving problems and building great products.

Imagine you are trying to decide between two equally important features to present to the leadership team for buy-in. Before you do that, you share these two features with your tech teams to get some early estimates. After discussing with your developers, you realize that one of the features will take one day to build vs. another that would take three months. Wouldn't your pitch to leadership change based on this newly acquired information? Engineers can often make these quick estimations, and you should leverage their instincts in picking ideas that give the most return on investment.

You could define the biggest, scariest problems your organization faces and do a "call for solutions." While not every solution might be worth pursuing, by framing the largest problems, you give your team members the opportunity to think about and empathize about the same thing. This can also save you from the trap of falling in love with your own idea. Great ideas can come from anywhere. As designers and managers, our goal is to let the best idea win, not the idea of the person with the best compensation.

Don't dictate what tech teams should build. Don't tell them what code to write or how to write it. Start with simple questions like "Hey, wouldn't it be cool if we could do X?", "Hey, is there some way to build this?", or "What's the fastest/best way to make X possible?" Collaborate with them

---

[8] Lawson, Jeff, and Eric Ries. *Ask Your Developer*. Harper Business, 2021.

and trust them to make great engineering decisions. Don't turn to your engineers just for code, but also for creative problem-solving. When you tell someone exactly how they should do something, it's a one-way street, and you don't engage them. But when you ask someone to solve a problem, the conversation becomes more collaborative, and they are likely to become more engaged.

Motivation

Motivated and engaged teams are key to effective collaboration. When you're a designer working in an AI product team, motivating and engaging your team members is less about standing in front of a podium and giving a sermon and more about small, regular, and consistent interactions about the user experience. While you might sometimes get a chance to present in front of a large audience, you would often communicate with a small set of people in person, over video calls, through documents, and over email or messaging. Don't wait for the perfect solution. Share progress with your team members and involve them early. Inspire them with examples, slideshows, personal stories, vision videos, prototypes, or even analysis from user research. Help them build a deeper understanding of your product principles and experience goals. Aim to create a feeling of urgency, energy, and that what they are building is important. Here are some ways you can engage and motivate your team members.

Build User Empathy

Make it a habit of thinking about the user experience all the time within your team. Always tie your problem and solution to the user. If presenting to a new audience, describe your persona and the user's journey. While it can feel unnecessary to show UX artifacts to engineers, I can't overstate the benefits of doing this. Enabling your team to understand user needs deeply and then letting them meet them is what sharing problems is all about. As product designers and managers, you have the power to connect your team members with customer needs. Great designers and product managers are not a barrier between customers and product teams. In fact, they remove barriers, eliminate preconceived solutions and erroneous presumptions, and streamline communications. They facilitate the understanding of the customer problem. Once a team member has built empathy for the user, the act of writing code or building a workflow becomes trivial.

You want your team members to constantly ask questions about users and how they use the product. You want them to get a feel for what's real and start forming a picture of the user. The following are some techniques you can use to build user empathy:

1.   Show a day in the life of a user.

2.   Present video or audio snippets from user interviews.

3.   Show snippets of users interacting with your product.

4.   Share documents and photos of your user persona.

A customer-centric organization is one that constantly self-corrects to put customers at the center of decisions. By building empathy for the user, you resist the urge within your team to make decisions that deprioritize customers.

> When you're a designer working in an AI product team, motivating and engaging your team members is less about standing in front of a podium and giving a sermon and more about small, regular, and consistent interactions about the user experience.

Transparency About Product Metrics and User Feedback

Show your product's progress, positive and negative customer feedback, and relevant metrics like customer funnels, product usage, etc. with your teams. Be transparent about what's working and what's not. Explain your AI product's goals and how it performs against those goals. For example, if your goal was to improve engagement, explain the improvement in engagement through usage metrics before and after the AI was deployed.

And even when things don't work, conduct a blameless postmortem. The purpose of the blameless postmortem is to dig below the surface of some kind of bad outcome to the true root cause and address that as an organization.[9] When things go wrong, it's either a time to blame or a time to learn.

Storytelling

Once your team members understand whom they are building for, you can present your vision and solutions through various user-centric narratives like user journeys, videos, slides, or storyboards. You can also present your product's before and after stages and show how the change improved the user's experience. Pair your stories with personas and metrics. Your stories and narratives should not be specifications, but instead, they should be an input in the product development process to drive better discussions about what and how to build.

Encourage Experimentation

Building robust AI models requires a lot of trial and error. Many times, it is harder to figure out the right thing to build than actually to build it.

---

[9] Lawson, Jeff, and Eric Ries. *Ask Your Developer*. Harper Business, 2021.

When trying to find the best possible solution to a problem, rapid iteration, experimentation, and close contact with customers are critical. You can help your team members get comfortable with iterations by telling stories, showing prototypes, and empathizing with users. Try things in a low-risk way and quickly learn about your customers' needs. Encouraging experimentation is good for your ML pipeline's robustness and your ability to influence the product effectively.

Hypothesis Validation

Experimentation is critical to innovation. There will be failures. To encourage and enable experimentation, you need to build a tolerance for failure. An excellent way to build tolerance is by reframing failure as learning. Instead of starting with a potential solution, start with a problem and a hypothesis. For example, making the feedback button larger might be a possible solution to the problem of low feedback that fails or succeeds. However, framing this problem and solution as "improving the size of the feedback button will increase the quantity of feedback by X%" is a hypothesis that can be validated or invalidated.

When you say that the experiment failed, it can feel negative, placing the blame on a person or the team, and could even discourage future experiments. On the other hand, when you say our hypothesis was invalidated, it is blameless; it feels like we learned something, motivating us to move forward.

Gathering Better Functional Requirements

When working on AI products and designing experiences for users, you also need to understand the requirements of various internal stakeholders. These can be in the form of workflows and mechanisms needed to collect data, feedback, limitations, and errors.

Data Requirements

Understand how data is acquired, collected, and generated. Most AI systems require lots of data to train. While data requirements can be different for different AI solutions, the following are some considerations you can start with:

1. Work with your machine learning teams to translate user needs into data needs. This can help you identify specifications for the type and quality of data needed to build a robust model.

2. Consider if you need to build mechanisms for collecting user data. Consider how you might communicate this to users.

3. In many applications of AI, there is a lot of reliance on labeled data. Consider if you need to design solutions to enable data labeling.

4. For supervised learning, having accurate data labels is crucial to getting useful output from your model. Thoughtful design of labeler instructions and UI flows will help yield better-quality labels and, therefore, better output.[10]

5. Understand how data is sourced if there are any privacy or ethical implications like bias in the dataset.

6. If you are collecting user data, understand what user consent and privacy implications exist. Check if there are any risks of revealing user data.

---

[10] pair.withgoogle.com, `https://pair.withgoogle.com/`.

Feedback Mechanisms

AI systems are not perfect; they are probabilistic and can sometimes make mistakes. They adapt and change over time. This change happens by correcting the AI's mistakes and identifying places where it can improve. The ability to learn is a key component of modern AI.

Learning in an AI system happens by providing it with data and giving feedback on its outputs. Feedback loops are critical for any AI system. Work with your team to understand different types of feedback needed to improve the model. Ask your team if they would need workflows for different implicit, explicit, or dual feedback mechanisms. Understand how feedback will be used and appropriately communicate to users.

Understand Limitations

Understand what your AI system can and can't do and when it works well and when it fails. Errors in your AI system are inevitable, and you need ways to handle them. Your users will test your product in ways you can't foresee during the development process.[11] Misunderstandings, ambiguity, frustration, and mistakes will happen. You will encounter different types of errors in your AI product. Some will be caused by faults in the system or data, while some might be due to the user's mistake. Sometimes users might presume something is an error, while your AI thinks otherwise. It can be challenging to find the source of an AI error. So you need to collaborate with your team to find out different errors and why they occur and figure out how to handle them.

---

[11] pair.withgoogle.com, https://pair.withgoogle.com/.

Highlight Ethical Implications

Ignoring ethical implications can erode user trust.

Encourage people to talk about ethical implications if they arise. Try to incentivize team members to raise ethical concerns. Those who advocate for ethical design within a company should be seen as innovators seeking the best outcomes for the company, end users, and society.[12] There are no prescribed models for incorporating ethical design in products. Leaders can facilitate that mindset by promoting an organizational structure that supports the integration of dialogue about ethics throughout product life cycles.[13]

Existing product development processes can be good opportunities to get your team thinking about the ethical implications of your system. The transition points between discovery, prototyping, release, and revisions are natural contexts for conducting such reviews.[14] Your team members can highlight concerns, identify risks, raise red flags, or propose alternatives in these reviews.

Raising ethical concerns can feel like an impediment to a project. Most companies have a goal of improving profits, market share, or shareholder value. Sometimes this focus on growth can come at the expense of ethical consequences. Ethics can take a back seat, especially in fast-growing industries and companies. However, ignoring ethical implications can erode user trust in your AI. Sometimes, you might be required by law to consider ethical implications. An AI that is not trustworthy is not useful.

---

[12] Ethically Aligned Design: A Vision for Prioritizing Human Well-Being with Autonomous and Intelligent Systems. First ed., IEEE, 2019.

[13] Ethically Aligned Design—A Vision For Prioritizing Human Well-Being With Autonomous And Intelligent Systems. 1st ed., IEEE, 2019.

[14] Ethically Aligned Design: A Vision for Prioritizing Human Well-Being with Autonomous and Intelligent Systems. First ed., IEEE, 2019.

Summary

This chapter discussed different roles in an AI team and how designers can work effectively with their machine learning and engineering counterparts. Here are some important points:

1. Building great AI products is teamwork. Different members of your AI product team need to work together to create a great user experience.

2. Start by understanding the interaction between you and your engineering team. Think of engineers less as people who churn out code and more like creative problem-solvers.

3. The complexity and the number of decisions required to create great AI products necessitate multiple individuals working in teams. The following are the most common roles in an AI product team:

    a. ML engineer

    b. ML researcher

    c. Applied ML scientist

    d. Software engineer

    e. Data engineer

    f. Data scientist

    g. Product manager

    h. Product designer

4. Training an AI model is a slow and iterative process. Engineers often need to use their judgment and imagination when tuning the algorithm.

5. Inspire your team members with stories, prototypes, customer feedback, and findings from user research and introduce them to UX methods and design principles.

6. When you're a designer working in an AI product team, motivating and engaging your team members is less about standing in front of a podium and giving a sermon and more about small, regular, and consistent interactions about the user experience.

7. The key to getting product teams and developers to work well together is for the product teams to share problems, not solutions.

8. Don't dictate what tech teams should build. Don't tell them what code to write or how to write it. Collaborate with them and trust them to make great engineering decisions.

9. Don't wait for the perfect solution. Share progress with your team members and involve them early.

10. When trying to find the best possible solution to a problem, rapid iteration, experimentation, and close contact with customers are critical. You can help your team members get comfortable with iterations by telling stories, showing prototypes, and empathizing with users.

11. Encouraging experimentation is good for your ML
     pipeline's robustness and your ability to influence
     the product effectively.

12. To encourage and enable experimentation, you
     need to build a tolerance for failure within your
     team. An excellent way to build this is by reframing
     failure as learning.

13. Instead of starting with a potential solution, start
     with a problem and a hypothesis that can be
     validated or invalidated.

14. You also need to understand the requirements of
     various internal stakeholders, which can be in the
     form of workflows and mechanisms needed to
     collect data, feedback, limitations, and errors.

15. Encourage people to talk about ethical implications
     if they arise. Try to incentivize team members to
     raise ethical concerns.

Epilogue

I started writing this book based on a belief that artificial intelligence is important and inevitable. In many cases, AI is already here. And while I don't think AI will solve all our problems, it will solve some of them and help us solve others. Sometimes it might even lead to new problems, and I hope these new problems are better ones.

Access to well-designed, ethical, and human-centric AI products will be a net positive for the world. Writing this book has helped me become a better designer and team member, and I hope reading it does the same to you.

Many of you will go on to design the next wave of AI products. You will be writing the future, and I am optimistic about this future.

I wish you all the best!

Akshay Kore

© Akshay Kore 2022
A. Kore, *Designing Human-Centric AI Experiences*,
https://doi.org/10.1007/978-1-4842-8088-1

Contact Author

This book would not be complete if I didn't practice what I preach. Just like AI products improve over time through feedback, people do too, and so do books. If you have any feedback on the contents of this book or want to just say hi, I would love to talk to you. :)

Reach out to Akshay Kore

www.aiux.in

OR

Scan QR code for contact information

© Akshay Kore 2022
A. Kore, *Designing Human-Centric AI Experiences,*
https://doi.org/10.1007/978-1-4842-8088-1

Index

A

Accountability, AI ethics, 354–356
Accuracy, 36, 140, 429
Actuator, 418–420, 430
Advanced home theater
systems, 416
Affected users, AI system, 123
Aggregation, 340
AI capabilities
AI products, 408
classification and
categorization, 422
CV, 409–411
input-output mappings, 33–35
knowledge
representation, 422
natural language processing
(NLP), 412, 413
pattern recognition, 424–426
perception, motion planning
and control, 418–420
prediction, 420, 421
ranking, 421
recommendation, 423, 424
speech and audio
processing, 415–417
trust in, 116

AI ethics, 325, 326
accountability and
regulation, 354–356
beneficial AI, 358
beneficial machine, 359–361
bias, 330–338
black box model, 329
control problem, 358, 359
ethics-based design, 326, 327
explainable AI, 328, 329
human in the loop, 361, 362
Independent Review
Committees, 357, 358
law, 356
liability, 357
manipulation, 343–347
privacy and data
collection, 338–343
safe AI, 348, 349
security, 349–354
transparency, 329, 330
trustworthy AI, 328
AI explainability, 375
AI metrics
accuracy, 405
confusion matrix, 404
precision, 405, 406

© Akshay Kore 2022
A. Kore, *Designing Human-Centric AI Experiences*,
https://doi.org/10.1007/978-1-4842-8088-1

Printed in the United States
by Baker & Taylor Publisher Services

Printed in the United States
by Baker & Taylor Publisher Services